Olympia, 4.18.11

To Mick and J.B

con simpatia e

stima.

[illegible]

GIANNI MAIMERI

Treatise on Paint

EDITED BY SANDRO BARONI

UMBERTO ALLEMANDI & C.
TURIN ~ LONDON ~ VENICE ~ NEW YORK

I have been working for the Maimeri industry without interruption since 1945. I worked in thc laboratory alongside my father, Gianni Maimeri, until 1951. My father died unexpectedly on November 25, 1951, and at the age of twenty-five I suddenly found myself alone at the helm of the company. I still had a lot of questions that remained unanswered. But everything my father had taught me encouraged me to continue in his path, and, I hope, in his style.

When I was young, many of my friends came to visit me in the evenings, not so much because they liked me, but because they liked my father, who had a lot to teach them on all kinds of subjects.

My father's *Diaries*, published in 1996, his *Treatise on Paint* and his *Treatise on Colour*, which he wrote in 1933, are the written confirmation of everything that I learned directly from him.

Leone Maimeri

LEONE MAIMERI
President Maimeri spa

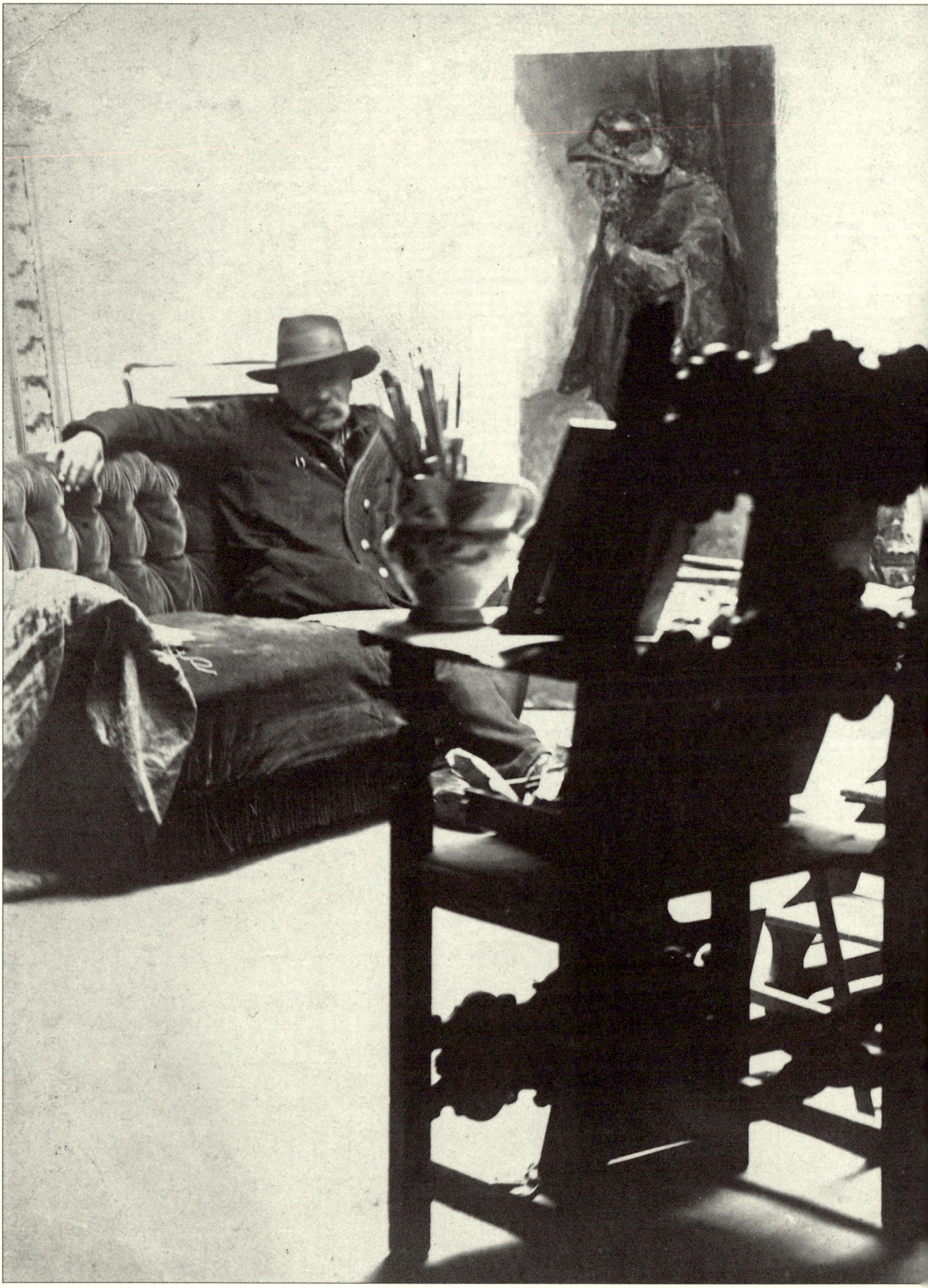

Gianni Maimeri in his Studio, Barona department.

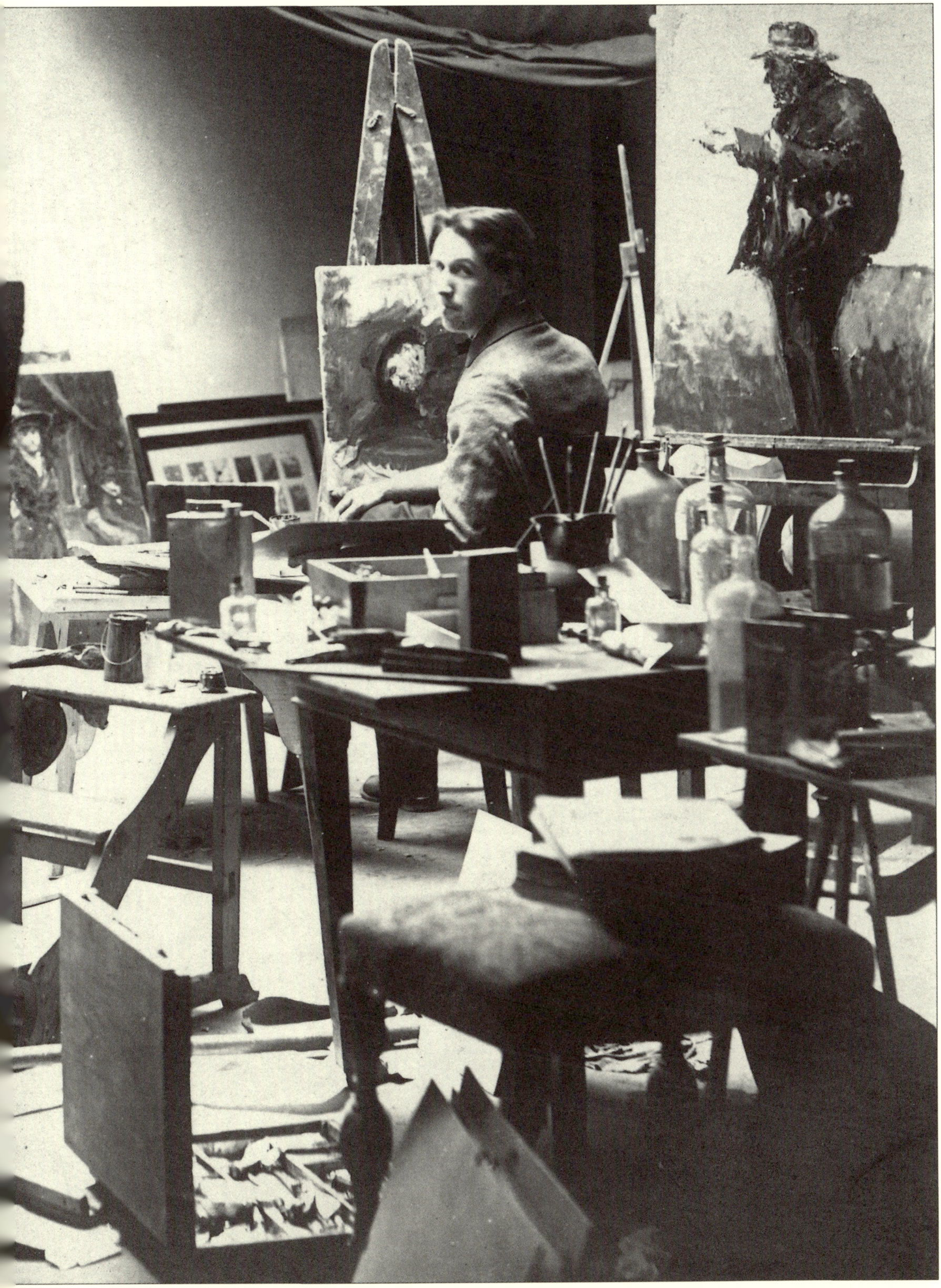

Contents

Introduction

SANDRO BARONI

One always feels somewhat ill at ease when editing texts or documents an author never had printed.

This is not today so much the product of respect for what we might call the author's "private" space. The assault of the media and the all-seeing eye of Big Brother have habituated us to a more elastic definition of the separation of the public and private spheres. My doubts are more likely to arise from the weight of the responsibility for making into a "historical document" something that the author certainly would not have intended to become so "historical". Or rather, historical in this form. Flaws in syntax and grammar, oversights, slips of the pen, drafts and thoughts which were suspended and never fully developed are handed over to the public in a posthumous edition which is treated as scientific, and I can already see myself in a dream, or perhaps in another time, in one of those never-ending and, one should hope, peaceful expanses of the plains of the great beyond, being questioned by Gianni Maimeri, with his beautiful colour palette, which I have just finished restoring, about why I didn't delete this or that repetition, rephrase an awkward sentence, or leave uncorrected a tiny grammar mistake which is of course merely an oversight, but still doesn't cast him in a good light.

Dear Gianni, I worked as a restorer works: throughout my career, I have never sought to alter an original with useless retouching. Now that I concern myself with historical texts on technique in the arts, little has changed for me.

"Beauty is the splendour of truth". This is what Thomas Aquinas said, and of course I still believe it. You thought the same thing, for you always portrayed the truth, seeking the filter of poetry in it: the beautiful. I think it, or, more often, I have trouble thinking it, because the beautiful and the true are sometimes disturbing, painful, more often imperfect, with their way of referring us to the "Other". The Greeks knew this, for they portrayed Aphrodite as cross-eyed.

But it is precisely this imperfection, this fragility of drafting, hinting at things, of suspended thoughts, of oversights are what we find most fascinating. And they fascinate us with the reprinting and reading of this apparently naïve little volume on painting technique. Here, for example, in the double introduction, present in the first draft and then in a "revised and corrected" version, appears the right intolerance of pedantry, which still today strikes us and puts a smile on the face of the experienced person who finds himself having to work with technical treatises: first "a rehashing of nonsense" and then of "clichés and many inaccuracies".

The beauty of truth. First naked, and then "benignly dressed in humility".
For these reasons, as well as scientific reasons and ones of personal inclination, I have sought and opted for a method of editing the text which could leave space for its own originality and temporary nature, in view of the author's many "variants". But all this in a form which could still permit a "straightforward", "cursive" reading of the work, reading only the text and ignoring the footnotes.
A choice which I consider - though I hope I am wrong - perhaps rather bold today, in view of the increasingly widespread trend towards popular science, from "the bottom up", that is, from the point of view of a person who has nothing to say, but wants everything he doesn't know, and needs to learn, to be reduced to suit the modest parameters of his own cultural and mental status.

But let us come to the presentation of this work, that is, to a number of thoughts and considerations on its significance, yesterday and today.
First of all, Gianni Maimeri's *Treatise on Paint* is a work which the author never finally revised for publication but which is almost complete, in the sense that it has a preface and a close, and which comes down to us, as a draft, in the form of a manuscript in the author's own hand and a few typewritten pages.
It was almost certainly begun and mostly written in the early 1930s, and completed by 1946-1947.[1]
It was most likely the general economic circumstances, connected with the war, and Maimeri's other personal and business commitments that prevented him from publishing the text.
To understand the little *Treatise on Paint* in full, it is definitely important to understand the painter Maimeri's relationship with artistic techniques: that is, what role painting techniques played in the author's ideas, especially with regard to his production of paintings.
The particular personality of the person we are considering here as the writer of the *Treatise on Paint* will provide some guidance for the purposes of preliminary orientation.
Gianni Maimeri had a good general education. He was a man with a range of highly personal interests, many of which he cultivated and studied in depth. While he was constantly busy with many commitments and self-taught in a number of fields, he was capable of rapidly gaining an overview of the state of knowledge at the time on any particular issue. This was thanks to his typically Lombard pragmatism, which allowed him to approach study of cultural issues of great complexity without any prejudices or sense of inferiority. With these he combined a sort of modesty which was rooted in his profound respect, and, we might say, love of all forms of human knowledge.
He was gifted with intelligence and curiosity; for curiosity is perhaps one of the clearest

signs of intelligence; he was never superficial, but was also capable of avoiding excessive doubts and pedantic distinctions.

He had received a good classical and humanistic education, read Latin, spoke French perfectly and also spoke English and German fairly well.

He was fascinated by the natural sciences, and by botany in particular: he had a herbarium of dried plants, samples he had collected over the years, kept in a dozen or so folders, along with a few thousand preparations, all duly classified and ordered.

He had a good knowledge of music, played the piano quite well and also seems to have played the guitar. His elder daughter Zoe was an internationally renowned concert pianist. He assiduously attended concerts and his season's tickets to La Scala inspired a series of drawings of "Musicians" offering a highly personal interpretation of the relationship between music and graphic representation.

The catalogue of his personal library, dating back to the 1930s and still kept by the family today, includes works of various different kinds, including some antique editions which had probably been handed down in the family. These included a good number of Latin classics, encyclopaedias and works on the natural sciences, as well as technical texts and works of social interest.

This eclectic personality and the practical, pragmatic spirit we have mentioned corresponded to a versatile appreciation of practically all the techniques of painting in his artistic expression.

Having chosen painting, following a personal inclination or vocation which appeared early in his life[2] (Maimeri almost always called himself a "painter", very rarely an "artist"), techniques were the "technical means" permitting expression of his "personal talent", means which had to guarantee, as far as possible, that the work "would be maintained unaltered" according to "the artist's imagination" destined "for creation of a new expression". Means which however "present their own prerogatives", and in this also "limitations". The important thing is not to seek in given techniques "results that cannot logically be achieved", such as the case of someone desiring "matt oil paint or glossy pastels".[3]

Gianni Maimeri therefore sees painting techniques as an idiom, a "means", for expression of the artist's "imagination", means which however have individual characteristics which are "logically" their own, and which the artist chooses consciously, in view of his own goals. In this there may be an analogy with music, in which a composer may choose which instrument in the orchestra he is going to ask to perform a series of measures on the basis of the instrument's real acoustic possibilities and his own expressive intent. Unlike music and instruments, however, the teachings of painting techniques must be "critically sieved", for each artist has, according to Maimeri, one great concern: "the desire that the work will be preserved unaltered in the form he wants it to have".

And now we may ask to what kind of literature we might compare the unpublished *Treatise on Paint.*
In other words, it is important to understand, precisely for the purposes of better comprehension and historical assessment, what the relationship is between the preparation of the work and the technical and artistic literature that comes before it or is contemporary to it. And how much of all this Gianni Maimeri was likely to be familiar with.
The artist wrote his treatise in the early 1930s, as we will see, almost certainly in relation to and as a reaction to other works being published in popular editions and in publications distributed by art supply shops and paint manufacturers at the time.
An understanding of the situation of technical and artistic treatises in Italy and in certain other European countries will therefore help us put into context and better understand the conditions and choices that led the author to write the manuscript in the form in which it comes down to us.
We cannot of course go over all the technical literature which had, at least since the mid-nineteenth century, produced profound changes in the orientation of transmission and reflection on artistic techniques here. But we will go over some of the most important tendencies.
The progressive and increasingly broad evolution of sciences such as chemistry had, in the nineteenth century, radically overturned all those expositions and transmissions of technical and artistic knowledge that were based on or drew their authority from tradition or from an artist and treaty writer's direct empirical and personal experience. In many cases, moreover, the new discipline became the very instrument by which tradition could be assessed and evaluated, and in this, observing the best preserved works, it seemed possible to find the key to the general and structural principles of optimal execution through chemistry.
The birth of restoration as a profession and a discipline in its own right separate from that of the painter, or at most the retoucher of the earlier tradition, had on the other hand helped lay solid foundations for the idea that long successful preservation of the works of antiquity depended essentially and almost exclusively on the use of good technique in their execution.
Previati himself, whom Maimeri considered one of the cornerstones of treatise-writing on technique in the arts[4], wrote at the beginning of his treatise: "Painting techniques embrace the practices necessary to give paintings consistency and durability...".
"Consistency and durability" are the themes we will see reappear, as an undercurrent, in Maimeri's writings too.
The scientific approach had therefore unhinged the inertia of tradition and restoration had renewed the focus on the role of technique as a source of stability in the conservation of paintings. But another factor was appearing on the horizon, at different times in dif-

ferent European countries, with an impact on the general shift of focus in the transmission of artistic techniques: industrialisation.

The full success of industrially produced paints for the fine arts, with widespread distribution of products which were increasingly becoming available to the new social classes that had already been present in the past but were becoming more influential both numerically and economically, brought with it a need to provide information and to promote and disclose proceedings and techniques which had previously been the exclusive concern of professionals or at most erudite amateurs, lovers and appreciators of Apelles' art.

The result was the creation, or rather establishment, starting in the second half of the nineteenth century of two different types of publication which, at least in part, corresponded to different types of users and uses. On one hand there were bulky treatises of an erudite academic type, the heirs of a long tradition of art treatises: works which were necessarily long-winded, often monumental in length, which Maimeri referred to as "ponderous volumes" ("I have one in my library consisting of ten volumes plus illustrations"). On the other hand were the modest little volumes, increasingly numerous since the early years of the twentieth century, which were limited in scope but often essential and capable of going straight to the heart of the matter and responding to the most practical requirements, and not so interested in erudite historical developments; manuals which were read by a plethora of amateur and professional painters, mainly from the bourgeoisie, who were definitely more interested in pragmatic concerns and positive information than in intricate, tired historical and academic disquisitions into the techniques employed by the ancients.

For a number of reasons, in other European countries and in the United States, works of the latter kind, which actually drew on earlier traditions in many ways, had met with fortune and success among a larger public since the start of the twentieth century, if not since the start of the Belle Époque. In Italy the full flowering of this type of publication did not come until later, with respect to the publications in French and English; but it was copious and rich in episodes of significance, with implications that remain to be studied. In the three decades between the end of World War I and the early 1930s, a dozen or so little volumes of this kind were published, primarily by artists and/or manufacturers or technicians of the manufacture of paints. If we look at the first three decades of the twentieth century and broaden our scope to include drawing and varnishes, the number of publications almost triples.[5]

Among painters, after Previati (who however wrote a work which may be classified in the first category described), we find Giuseppe Ronchetti, Califano Mundo, Carlo Linzi, Dardo Battaglini, Gianni Maimeri himself and, finally, Giorgio De Chirico, as well as engravers such as Giulio Cisari. Manufacturers and manufacturing technicians who authored such publications, only too often anonymously, publishing works which did not cite their author, included Giovanni Sommaruga, Max Mayer and Pietro Bonomi Da Monte, respectively manager and director of Colorificio Italiano, and once again Gianni Maimeri. The two types of works were largely distributed separately: the academic ones in the circuit of the big publishing houses and the more traditional forms of distribution through bookshops and academic institutions; the "handbooks" on the other hand were published as part of inexpensive popular series such as the "People's Library" published by Sonzogno or the same publisher's collection of "Practical Handbooks", Salani's "Library for All" and the unmatched series of manuals which Ulrico Hoepli began publishing as early as the final decades of the nineteenth century. The majority of these editions were sold primarily by correspondence. Orders were placed directly with the publisher, specifying the title and sending payment of the price specified in the catalogue, and the requested publication was sent in the mail.

But by the mid 1920s, greater business interests in a growing market made different solutions available.

On one hand, companies with strong roots in tradition promoted *causeries*, true public conferences, often held in prestigious locations, in which a public of selected artists was invited to discuss and debate a technical-artistic issue. This was followed or marked by publication, sometimes in more than one language (English and/or French) which was distributed free of charge through the network of agents and salesmen, publicising the content of the communications. Thus an event was created and followed by distribution of a technical publication related to the topics discussed, which conveyed the manufacturer's technical choices and know-how, in however discrete and general terms. For example, in this way, on 24 May 1928 Talens & Zoon promoted a meeting in London with a public of artists, in which they discussed colours, oil paints and varnishes for artists and the techniques of the old masters. This was followed by publication of a twenty-page brochure, which I am familiar with in the French edition distributed free of charge by Apeldoorn. After careful consideration of the technique of the old masters, the last part of the text directly mentions the manufacturers of Rembrandt oil paints, and it was clear from a subtitle on the cover that Talens & Zoon were themselves the manufacturers.[6]

Raffaele Armando Califano-Mundo of Naples had a more innovative idea and one that was also more complex from the commercial point of view.

Towards the end of the second decade of the twentieth century this unusual artist, art lover, publisher and businessman had already established a solid publishing business in Porta Piccola, Capodimonte, Naples. The titles of his series, some of which were included in the "Artist's Library", ranged from *Aluminuria and Pastel Painting* to *Manual of Gouache and Tempera Painting*, *Manual of Encaustic and Fresco Painting*, *Manual of Enamel*

Painting, Manual of Painting on Glass, Manual for Tapestries, Gobelins, French Weaving, Dry Tapestries, Brocade, Treilles, Fougère, Manual of Varnishes and Varnish Paints, Manual of Drawing and Precepts. Numerous editions were published of the *Manual of Oil Painting, Manual of Colours* and *Manual of Watercolour Painting - Miniatures and Restoration.*[7]
Califano-Mundo's publishing projects were a sharp intuition. The cost of publishing the manuals he produced was largely paid for by advertisements for manufacturers of paints and materials for the fine arts, which appeared on the second, third and fourth pages, with inserts of up to four pages permitting further development of the material contained in the advertising. In the Manual of Oil Painting, for instance, the "sponsors" include several of the most prestigious makers of fine arts products: the Pietro Miliani paper mill in Fabriano, Talens & Zoon, J. Blockx Fils, Schoenfeld & Co., Coloreria San Carlo or Paciosi of Rome, Schmincke & Co., Kaspar & Co. of Vienna. Midway through the little book of four groups of sixteen pages was a collection of colour samples for Mussini extra fine oil paints. The volumes were sold by the artist and publisher by correspondence, but were also distributed by a number of Italy's principal paint manufacturers, especially in central and southern Italy.
Gianni Maimeri was definitely aware of the flourishing of these publications. One of his main Roman distributors, Paciosi, whom we have already mentioned, had close business relations with Califano-Mundo, and in fact in the advertising that came with the *Manual of Oil Painting*, the many specialities advertised in Paciosi's full-page ads include "Maineri colours"; clearly a misprint.
There are too many analogies between the two works.
We must point out that Maimeri, who spoke French perfectly, was certainly also aware of the promotions and *causerie* of his major competitor Talens & Zoon. And so it might occur to the artist and entrepreneur to provide a mediation summing up the situation and, at the same time, offering his own guidelines, his own experience and, why not, his own products. Maimeri wrote in his Closing: "And so I must ask the reader's pardon if I have often had to make reference to a manufacturer in order to describe products, and the suspicion may be that I wish to advertise them. Certainly, several manufacturers, including a number of Italian ones, produce excellent materials... But I can guarantee the one I have mentioned, because I myself have directed its technical management ever since it was established, and many of its products are precisely the fruit of the meditations I have outlined above for my colleagues in the hopes of being of assistance to them."

At this point we must note the close relationship between the cultural and technical horizons of Maimeri's brief exposition and the world of restoration.
The very concepts of "inalterability and stability" which Maimeri puts at the founda-

tions of his exposition are indirectly taken from the science of conservation, then in its infancy.
The best-known nineteenth-century Italian restorer, Giovanni Secco Suardo, is explicitly mentioned in the text twice: first as the "founder of the art of reasonable, scientific restoration", and then as being "considered a master ... of technique".[8]
Maimeri also considers Previati[9] one of the great essayists, "though imprecise", for he wrote and contributed an interesting appendix to the book *The Restorer of Paintings* by the Bergamo count and restorer. This essay places much emphasis on the theoretical foundations of scientific restoration, which was gaining ground precisely at that time in Italy. Maimeri would certainly have read this text and been familiar with it. The edition of Secco Suardo's manual in his possession was in fact the 1918 edition, containing a lengthy note by Previati.
All of the section of the treatise concerning varnish paints is written with both painters and restorers in mind.
The formulation of varnish paints, later referred to as paints "for restoration", in the black label line was the first in Italy and is still practically the only industrially produced line of this kind in the world. It is only understandable that Maimeri should dedicate so much space to it in his brief treatise, aware as he was of the potential of such an unusual product line.
But we know from other sources how close the bond between the world of restoration and other product lines initiated by Gianni Maimeri was, and what its implications were. The line of paints we have just mentioned was perfected by directly consulting a restorer (...), who later moved to Paris to work at the Louvre.
Our treatise also mentions earths such as Bergamo earths (particularly a yellow ochre). Until the 1950s the supplier and procurer of a few truckloads of this valuable product for Maimeri was one of Italy's most famous restorers, Mauro Pelliccioli, destined for great fame and well-deserved prestige after the war.
Those were years in which restorers, though already recognised as a class distinct from painters, were still familiar with the technique and practice of painting, in both the executive proceedings and the concrete study of materials. This knowledge, in most cases real and not a pretence, permitted profitable exchanges of real value and effective synergies with other skilled craftsmen concerned with paintings: painters, decorators, art historians and makers of materials for the fine arts, as in our case.

While we have just touched upon some of the principal circumstances that might best explain the significance of the *Treatise on Paint* at the time it was written, in a generic "yesterday", let us now focus on the significance of its publication today.
I remain convinced that historical studies in general are not and must not be exclusive-

ly a matter of reconstruction and explanation of the circumstances and causes of events of the past alone, but have the implicit purpose of opening up the path to reflection about what might be the message the text can still pass down to our contemporary sensibility. Whether or not we like it, of course, in the present day, even this little, fragile, incomplete work - "naïve," some might say - still testifies to the fact that there was a time, not so long ago, when painters concerned themselves with painting techniques, and please note that this is "techniques" in the plural, and not just "their own technique". There was a profound relationship tying together everything that contributed to the physical creation of the work as an object of perception: the proceedings of knowledge and handiwork, the product, the processing of and way of using materials and tools of use to and functional for the genesis of the work.[10] Painters were growing in this area, as well as in the practice of painting.
Maimeri wrote in one of the passages that best, and I think best of all, explain the "painter's eye": "to the painter there are no more coloured objects, but results of the compositions of pastes. What is to you a lovely flower, the subject of a notion or the source of a sentiment, is to me a harmony of pink lake, pale blue, white and light yellow, cooled down a little by emerald. While remaining a flower, even so".[11]
In-depth knowledge of materials and therefore of their potential for the purposes of expression.
Once again, the desire to make something that will not be ephemeral, in which striving to achieve skill and quality contributes to expression, as well as to long life.
To the contemporary eye and judgement, especially "artistic", or "art-historical", these simple guidelines may seem truly naïve. But they are those of the entire tradition of painting in relation to techniques; with this we truly may ask ourselves, if we should stray from it, whether it might not be our time, or our position, that prejudices us, and whether freedom might not lie elsewhere, perhaps among the different forms taken by a constant of history.

[1] The manuscript was most likely written in a fairly short time, at least the first half of it, almost "straight off", as would seem to be demonstrated by a number of clues pertaining to the writing instruments used, such as the type of paper and pen. The majority of the text was written in no more than fifteen work sessions, as demonstrated by the clear resumption of "sequences", with only modest changes in the handwriting, changes in the inclination and ductus of the letters, pressure, ink and separating lines. The draft was, however, reread, corrected and modified repeatedly by the author, on more than one occasion, as demonstrated by deletions made in different types of ink and the two surviving versions of the brief introduction. The majority of the text was probably composed not long after the *Treatise on the Study of Colour*, an important unpublished work of a philosophical, phenomenological nature which was written mostly in 1932 and 1933. The paper is the same, taken from a single notepad, and the pen would also seem to be the same; there are four sheets of legal paper with the watermark of the Kingdom of Italy. In any case, the second half of the work includes notes on the production of gouache colours and, above all, varnish colours which may be dated around 1946-1947.

[2] Cf. *I diari di Gianni Maimeri*, G. BUCCELLATI ed., Milan 1996.

[3] Cf. *Treatise on Paint*, preface, para. 2.
[4] Cf. G. PREVIATI, *Memorie sulla tecnica dei dipinti* (1896); *La tecnica della pittura* (Turin 1905); *Della pittura, tecnica e arte* (Turin 1913).
[5] Cf., in this volume, P. TRAVAGLIO, *Italian Publications and Manuals on Painting Techniques* (1900-1940).
[6] The text of the brochure was largely based on Albert Abendschein's publication, *The Secret of the Old Masters*, published in New York in 1905, and drew on its title: *Le secret des Maîtres Anciens Causerie sur les coulores, huiles et vernis pour artistes et la technique des Maîtres Anciens*, Apeldoorn 1928. Similar promotional methods were not really so new. In the second half of the sixteenth century, doctor and writer Leonardo Fioravanti authored a very successful book of "Secrets" describing numerous preparations for colours and materials for the fine arts, in which he promoted the sale of a number of his miraculous preparations, available from certain spice merchants.
[7] A truly versatile writer, Califano-Mundo also wrote an *Encyclopaedia of Painting* in two volumes, a *Dictionary of the Greatest Artists from the Origins of Greek Art to 1930*, *The Evolution of Painting from its Origins to the Fall of Rome*, *History of Neapolitan Art since the 19th Century - The Landscape*, *Chemical and Physical Alterations of Paintings and General Rules for Avoiding Them*, *Little Dictionary of Utensils and Substances Used in Painting*, a *Treatise on Anatomy*, and a volume of *Critical Studies*.
[8] Giovanni Secco Suardo (1798-1873) is fully entitled to recognition as one of the most important players in the definition of modern restoration, in particular of paintings, in Italy. Various essays introducing the issue provide an account of his work and his theoretical contributions (cf. A. CONTI, *La Lombardia del Secco Suardo*, in *Vicende e cultura del restauro*, in *Storia dell'arte italiana*, vol. III, Turin 1981, pp. 84-86; and Giovanni Secco Suardo, "La cultura del restauro tra tutela e conservazione dell'opera d'arte", deeds of the International Study Convention (Bergamo, March 9-11 1995), supplement to issue 98 (1996) of the *Bollettino d'Arte*, 1998). The Bergamo count and restorer was also however an expert on painting techniques, and by 1858 had authored, among other works, a note "On the discovery and introduction into Italy of the present method of oil painting" (Milan 1858). Just before his death he published his *Thoughts on Encausto, Oil and Gouache Painting* (Turin 1870). The Angelo Mai Library in Bergamo preserves a number of manuscripts he left to the civic library on his death. Recently surveyed as part of the PROMATART project conducted by Fondazione Maimeri with the support of the Region of Lombardy under the writer's direction, they also contain notes on artistic techniques, translations of manuscripts of foreign works and formulas or descriptions of restoration proceedings. The study of these documents following the survey is still underway, but we may identify manuscripts MM 434 (cf. E. PIANURA, *"Regole particolari per diversi colori da miniare" e altri trattati di miniatura nei manoscritti della biblioteca Mai di Bergamo (MM 434 e MMB 528) messi a confronto*, graduate thesis, University of Milan, 2007-2008, supervisor Silvia Bianca Tosatti), MMB 639 (Italian translation, probably in the author's hand, of *The Art of Fresco Painting* by M. P. Merrifield) and MM 195 (para. 11. "Treatise on the varnish commonly referred to as Chinese by Padre Filippo Bonanni of the Company of Jesus. Second edition. In Rome 1731 by Antonio de Rossi. This manuscript is extracted from the original in the existing print in the hand of NOB: SIG: CO: L: S: S:", to be identified as Leonino Secco Suardo, commissioner of the Carrara Academy in Bergamo) in Angelo Mai Library in Bergamo.
[9] Gaetano Previati (1852-1920), a painter active primarily in Lombardy, also pursued an intense career as an essayist. He came to his own personal interpretation of Divisionism and later became one of the principal members of the European Symbolist movement. Umberto Boccioni explicitly admitted that he saw him as a starting point for the pictorial techniques of Futurism. He authored numerous publications and popular technical manuals including *Painting Technique* (Turin 1905) and *The Scientific Principles of Divisionism* (Turin 1906).
[10] Here I paraphrase a passage from the introduction to S. BORDINI's work, *Materia e immagine. Fonti sulle tecniche della pittura*, Rome 1991.
[11] Cf. G. MAIMERI, *Studio sul colore*, unpublished manuscript in the author's own hand, Maimeri family archives.

Criteria Applied to Editing of the Text

SANDRO BARONI AND PAOLA TRAVAGLIO

Gianni Maimeri's *Treatise on Paint* is preserved in three separate versions, all kept in the Maimeri family's archives. These have been collected together in a single file folder bearing the title "Piano Regolatore di Milano - Scritti vari", including a press review and typewritten letters concerning the debate about closure of the inner circle of the Canals and, separated from this by a piece of card, "various writings" in Gianni Maimeri's own hand, listed on the cover:

- Treatise on paint
- Treatment of a study of colour
- Notes on colours
- Preface
- Extraordinary meeting of the union

The three versions of the *Treatise on Paint* text that have come down to us are:

- *Manuscript in the author's own hand*: written on both sides of twenty-one 295 x 210 mm pages. Papers 9-23; 28-29 have thirty-one lines prepared for printing in grey ink. Papers 24-27, on the other hand, consist of two double sheets of "legal paper" of twenty-five lines each, also grey, cut in half. These papers have large watermarks printed with the coat of arms of the Kingdom of Italy and the name of the Ambrogio Binda manufacture, respectively. The manuscript is missing the first pages. The papers are loose and unbound, but partially numbered in Arabic numerals (in pencil, at the top right, and in red ink, at the top left) and bear frequent marks referring the reader from one page to another in the lower right and upper left corners. Excluding the above-mentioned sheets 24-27, there are two other types of paper: the first (cc. 9-23) are of substantial weight, which is not true of the similar papers cc. 28-29, decidedly less heavy and used in part to write out the other texts referred to above, contained in the same file. Both have no watermark and are of poor quality, and are probably of domestic manufacture. The writing is in italics with two different pens: a blue ink pen with a fairly fine line (up to c. 14v., r. 1) and a black stylographic pen with a wider, heavier line (from c. 14v. to the end of the manuscript), substantially using all the space available for writing. The same black stylographic pen has been used to make corrections and additions to the first part of the manuscript.

- *Typescript A*: typed with a mechanical typewriter on the back of eight numbered unbound 280 x 210 mm sheets of paper, with the watermark "Extraforte C.M. Fabriano" (Cartiere Miliani Fabriano) of about thirty lines each, with blue ink for the main text and red for the titles, written by the author's son Leone with the intention of establishing a text for publication and distribution of his father's work in paint manufactures. The text is interrupted after the words "fitta schiera di intrugli. La base di (...)" (c. 8). Even with the support of what must have been the first few pages of the original manuscript, destroyed after they were copied, this testimony reveals a number of evident interpolations, confirmed orally by the person who made them. These essentially regard updating of the colour palette to include pigments which came into use just after the war.

- *Typescript B*: typed with an electric typewriter on numbered unbound 297 x 210 mm sheets of paper, with about forty-five rows each, composed of seventeen sheets of paper written on one side only. This is a recent transcription, the author of which we were unable to identify, of typescript A and of the manuscript, put together and arranged so as to obtain a reconstruction of the text. It includes numerous variants which are clearly not original, as well as a number of misunderstandings or omissions identified by the copyist. It offers a useful draft transcription for comparing interpretations of Gianni Maimeri's highly personal handwriting, but is not of use for critical reconstruction of the text.

We have applied an editing criterion to the text and different versions of Gianni Maimeri's *Treatise on Paint* which we might define as semi-diplomatic. What we mean by this is that we stayed entirely faithful to the three surviving versions, including graphics, with perfect adherence to the form in which the text appears in the manuscript and in typescript A. For this reason, the numbering of the pages and row lines in the original are shown in the margin of the text.
The notes serve to make sense of the changes the author made during the task of preparing the text. Graphic marks are used to identify, among other things, the sole interpolation attributed to Leone Maimeri, key variations in interpretation appearing in typescript B and our own additions, which we have kept down to the minimum required for a complete understanding of the text. We have not, on the other hand, identified the division of words, upper- and lower-case letters, accents and apostrophes used to distinguish between homonyms, and punctuation on the basis of modern usage.
While these measures may to some appear superfluous, we believe this text to be of eminently historical interest applied to a literary document; moreover, putting together the

various versions reveals key stages in the preparation of the work by the author, as in the case of the preface. We therefore consider it necessary to apply scientific methods to make sense of it all, without thereby rendering the treatise unreadable for the general reader.

Legend:

< > additions

italics words which were underlined in the original

GIANNI MAIMERI

Treatise on Paint

EDITED BY SANDRO BARONI AND PAOLA TRAVAGLIO

(c.1r.) PREFACE

Yet another "treatise on paint"? The latest unwanted ostentation of petty erudition collected up with a rehashing of clichés and of the many inaccuracies handed down, almost as if from father to son, in the countless treatises, little manuals, handbooks? The usual mish-mash of quotations, from Pliny to Previati, with the usual praise of the expertise of the ancients and the decline of technique?

Far from it: this is intended to be the first attempt at objective analysis, at constructive critique of statements supported only by the weight of the writer's authority. A fine sieving (it gathers the fairest flower) of the formularies and recipe books to uproot the lies, blunders and idols that have, with tenacity and diligence, been administered since the days of Heraclius by the artifices of painting which, avid with specious secrets and attempting to find panaceas for their creative labour, rather than healthy assistance find in these books the complications and bungles that trap them in a swamp full of tribulations and tricks, from which they have difficulty extricating themselves in order to return to shore.

7 Caius ***Pliny*** *The Second:* Naturalis Historia. *We do not know with certainty which edition of Pliny's* Natural History *would have been available to Gianni Maimeri. It is likely that he referred to one of the two great editions widespread in the late nineteenth and early twentieth centuries: Datelfsen's edition (Berlin 1866) and Mayhoff's edition published by Teubner (1905-1909). It is possible that he was familiar with* Pliny: Chapters on the Hist. of Art *by K. Jex-Blake and E. Sellers (1896), and, around the time he completed his text, Silvio Ferri's* Storia delle arti antiche, *containing a translation of the passages on the arts, published 1946; Gaetano* ***Previati*** *(1852-1920), a painter who worked primarily in Lombardy, was also a prolific writer of treatises. He authored numerous publications and widely consulted technical manuals on painting such as* La tecnica della pittura *(Turin 1905) and* I principi scientifici del Divisionismo *(Turin 1906).*
11 ***sieving****: clearly a reference to the function of the Accademia della Crusca (literally* The Bran Academy, *an institution established to protect the purity of the Italian language) and its motto from Petrarch: "Il più bel fior ne coglie", "It gathers the fairest flower".*
14 ***Heraclius****: name or pseudonym of the author of* De coloribus et artibus romanorum, *probably known to Maimeri in Pellizzari's edition (1919), or through anthologies such as those by M.P. Merrifield (*Medieval and Renaissance Treatises on the Arts of Painting, *London 1849) or A. Ilg (*Quellenschriften..., *1873).*

We present here, to encourage brief reading, the original draft of the preface, deleted by the author, which survived in a copy of the manuscript in the author's own hand. (cc.19v.,r. 14 - 20v., r. 18) PREFACE/Yet another treatise on painting?/The latest ostentation of petty erudition/rehashing of nonsense handed down/from father to son, from treatise writer to editor?/[without] The usual author's notebook?/Far from it. The first attempt at an/objective analysis of the reality of painting/technique./A severe polishing off/(It gathers the fairest flower) of the formularies of/the prejudices and mistakes/which have since Heraclius' day been diligently/presented in good and bad faith/to painters hungry for the new/to return to the formula that will help them/relieve their torment and travail./ Many *(c.20r.)* may perhaps be disappointed to have to give up/idols they have long adored, perhaps in/silence. The result will be an honest though/brief list of truths which, if not mira-/culous, will at least be useful for those who do not want to/waste time but want to know/how to paint plainly and without /illusions./

Also, however, a useful interpretation of empirical observations which are left hanging in mid-air because they are not discussed in depth or supported by any systematic experimentation. Throughout my long career studying the problems involved in painting and varnishing technique, these vague indications have often led me to discoveries of ingenious practicality, pointing me down new paths for study.

I do not entertain any illusions that I will be universally praised by my colleagues. Many of them may be reticent to see the idols they have long worshipped torn down. The sieve is fine, and to tell the truth I have never been tempted to fill pages with print. I have resolved many synonyms together, reduced many different phenomena to a single proven cause. The worry of not making any statements that have not been confirmed caused me to give up many topics which are not yet ripe for discussion. The result is a thin little book which will not appear very impressive alongside the ponderous volumes of certain treatises (I have one in my library consisting of ten volumes plus illustrations). And yet it is the result, not only *(c.2r.)* of my personal experience making paint, but of attentive consultation of numerous valuable texts. While in scientific study we are not required to "jurare in verba magistri", it is no less true that everyone must bend down before the great masters, from Pliny to Heraclius and the monk Theophilus, to the precise Cennini, to Lomazzo and down to Mengs, to the long-winded Montabert al Mérimée, to our own Guareschi, to the acute Secco Suardo, founder of the art of reasonable, scientific restoration, to Vibert, the first to introduce the critical system to the highly technical aspects of painting, and even Previati, even though he was in many ways so imprecise and chaotic. Each of whom, along with infinite minor contributors, made a valuable contribution to the science of colours and paints, however mixed up with the usual chaos of repetitions translated without adequate critical judgement.

3 ***jurare in verba magistri****: Horace, Epist., I, 1, 14: "Swearing on the words of the masters".*
5 ***Theophilus****: name or pseudonym of the author of* De diversis artibus *or* Schedula diversarum artium, *a work composed in the first half of the 7th century, probably known to Maimeri in the edition published by R. Hendrie (1847) or A. Ilg (*Quellenschriften..., *1876); Cennino* ***Cennini****:* Il libro dell'arte, *a well-known treatise written in the late fourteenth or early fifteenth century, available to Maimeri in the edition published by Milanesi (1859), or, more likely, in that by Renzo Simi (1913).*
5-6 Giovanni Paolo ***Lomazzo****:* Trattato dell'arte della pittura *(Milan 1584),* Idea del tempio della pittura *(Milan 1590),* Della forma delle Muse *(Milan 1591); Anton Raphael* ***Mengs****:* Opere complete*: 1st edition edited by J.N. de Azara and printed by Bodoni (Parma 1780); 2nd edition edited by F. Milizia and published by Remondini (Bassano 1783); 3rd edition published by C. Fea (Rome 1787); Jacques Nicolas Paillot de* ***Montabert****:* Histoire de la peinture *(Paris 1829); Jean Francois* ***Mérimée****:* De la peinture à l'huile *(Paris 1830).*
7 Icilio ***Guareschi****:* Sui colori degli antichi *(Turin 1905); Giovanni* ***Secco Suardo****:* Sulla scoperta e introduzione in Italia dell'odierno sistema di dipingere ad olio *(Milan 1858),* Pensieri sulla pittura ad encausto, ad olio e a tempera *(Turin 1870).*
8 Jehan Georges ***Vibert****:* La science de la peinture *(Paris 1891).*

This does not mean that the enormous/amount of study on the part of worthy treatise writers is to be cast aside. Far from it/each of them has contributed a light and a/brick to the research work. And often/interpretation of some mistaken state-/ment by empirical researchers has/led me to valuable discoveries/in my long career studying the/technical proceedings of painting. Because/it is one thing to jurare in verba magistri, and another/to bow down before the trust masters/of art. With Pliny, when we manage to read between the lines of (...), Heraclius [Pliny] and the monk Theophilus/and/ponderous trea-tises, the great Cennini, Lomazzo/and even Vasari and right down to Mengs, to/Montalambert, to Merimee, to Guareschi/and to the perspicacious Secco Suardo and Vibert and/even down to Previati, who is imprecise in many areas/without mentioning the scientific treatises/which we will not discuss here, in all of these and/even in the anonymous little treatises in which/I have found something of use, some *(c.20v.)* precious confirmation of clear truths amidst/the usual chaos of repetitions compiled/and translated without judgement./Modest task with slim results./The sieve is very fine, and little remains/but at least it is quality material./The painters' greatest and/most justified concern is that of having/in hand a trustworthy product, as/unalterable as possible, which may meet their/need to express themselves clearly, with/readiness and without fear./But we will put off until the end/our [analysis] examination of the different ingredients in/paint with a little information on their [inaltera/bility] fastness and reciprocal/incompatibilities, attempting to provide a rapid/classification of paints./

To get right on topic, what is the greatest concern of the painters, and in what area can a manual be most useful? Every artist who has worked on creation of a new form of expression wishes that it may be preserved unaltered in the form his imagination desires, as he looks on with terror before the ruination of so many great works of the past, amidst which a few intact works stand out as islands. Finding the true reasons why some paintings are ruined while others are seemingly miraculously preserved will be the principal concern of this treatise. This study will allow us to formulate the preparations best suited to provide the painter with the technical media most appropriate to his own personal talent, teaching him not to seek results which cannot logically be achieved by certain techniques, as if one were to desire matt oil paint or glossy pastels. Each kind of paint has its prerogatives and its own limitations. And so let us put off to a later chapter the specific analysis of different materials used in the formulation of preparations for painting and start our treatise with a rational classification of paints.

(c.3r.) CLASSIFICATION OF PAINTS

In this classification scheme we will take into consideration only those paints which are suitable for executing artistic paintings, setting aside the infinite other varieties for industrial and artisanal use. Paints for artistic painting may therefore be divided into categories according to the way in which the coloured material (pigment) is bound so that it can be spread or applied to the surface used for this purpose (the substrate). We therefore have:

1) Paints in which the pigment is held together simply by cohesion, and is anchored to its substrate by adhesion. These are pastel paints, coloured pencils, chalks, and water-based mural paints (not fresco paints). They may also be called dry paints.

2) Paints in which the pigment is bound by a vehicle (medium), most often soluble in water, with agglutinating power (glue, rubber, etc.). These include water paints, tempera paints and gouaches.

3) Paints bound by a medium which is soluble not in water but in other solvents (resins, balsams, waxes, etc.). These include varnishes, encaustic paints, wax pencils, etc.

4) Paints in which the medium is chemically transformed to become insoluble to common solvents with time. These include oil paints, fatty gouaches and, in a class of their own, fresco paints or lime paints.

We must always keep in mind this rational classification, which is the basis of all possible techniques. The basic differences between the first three categories and the fourth category lie especially in the fact that in the first three, all or part of the medium disappears due to evaporation as the paint dries, while in the fourth category the medium, or most of it, remains and is transformed by a slow chemical process. And so the oil in oil paints becomes resin by oxidation, just like the fatty material in fatty gouaches, and slaked lime in fresco painting is transformed into insoluble calcium carbonate. This means that we may transform one of the paints in the first three categories into the fourth, by introducing (c.4r.) the elements characterising the fourth category into the paint, as we will see when discussing varnished gouache and other particular proceedings.

ON THE STABILITY OF DIFFERENT TYPES OF PAINT AND THE CAUSES OF DETERIORATION OF PAINTINGS

The inalterability of paintings has rightly been painters' greatest concern. And yet there are still very many prejudices and preventions regarding this very important topic.

Absolute inalterability is, of course, a myth. No material in the world is perpetually inalterable. But relative inalterability is possible, and the artist has two basic needs.

First of all, that the colours of the painting should retain their original hue and shade for as long as possible; secondarily, the paint should not deteriorate or be deformed from its original appearance. There are therefore two areas requiring study: one concerned with the materials included in the composition of colours, and the other with painting methods and preparation of colours. Let us look at the first problem first of all: the issue of inalterability of hue.

The hues of colours may be altered for four different reasons.

1) due to the effect of light on them.

2) due to exposure to natural agents (heat, damp, sulphurous emanations).

3) due to contact of coloured pigment with other pigments or with the materials making up the medium.

4) due to alteration of the medium of the paint, with which the pigment is surrounded or covered.

As we may see, in all but the last case, it is the pigment alone that is affected.

It is an abandoned prejudice and antiquated cliché of a few stubborn "laudator temporis acti" that the colours in old paints are superior to those in modern ones. While it is true that not all modern

34 ***laudator temporis acti****: Horace,* Arte poet.*, 173: "He who lodes the past".*

(c.5r.) makers of paint apply the diligence and scruple claimed by certain famous "Workshops" of the past, we can be sure that the wealth and variety of perfectly fast pigments available to any amateur scribbler today would have been the stuff of dreams for hardworking paint makers of former times, whose paint pots were full of countless colours now banned from the modern palette as poisonous. And is it true that those who know <or> are closest to knowing the nature of German blues and greens, of mountain blue, of blue ash risigallo, of the extracts of weld or the seeds of Persia, of liquorice juice, of yellow lac, of dragon's blood, of Gamboge, verzino and a thousand other colours, one more ephemeral than the other? For the modern painter has access to an almost indefinite variety of hues and colours which are perfectly lightfast and resistant to reagents in the atmosphere, with very few incompatibilities for mixing with one another. To cite only a short list of the principal colours unknown to the ancients, we might mention zinc and titanium white, cobalt blue, purple and green, emerald green and chrome oxide green, manganese purple and synthetic iron oxides, ultramarine blue <and the new very fast organic synthetic pigments such as wolfram, the blue and green phthalocyanines and the latest Cinquasia reds.>

The chemical action of light on colour is a well-known phenomenon and this is not the place to discuss its causes. Some colours disappear entirely or in part, others change colour, veering either to the right or to the left on the spectrum, while others definitely change colour and others yet become darker in hue. Of course the colours which change the most are to be rejected: while it may be tolerable for a green to lose some of its yellow pigmentation and become bluish, it is intolerable to see an indigo change to reddish brown. Changes in intonation are another pernicious defect, and I recall what a terrible and painful impression was made by Turner's marvellous sunsets, in which the evident darkening of chrome yellows, which had just come onto the market for the first time and were used by the great painter, caused the balance

18-20 An addition to the text present in typed copy A by Leone Maimeri.

(c.6r.) of light and dark hues in the painting to be completely reversed, so that the parts the painter clearly wanted to be lightest and brightest to appear dark.

A century or almost a century of experience working with these colours makes it easy for the scrupulous modern artist to compile a list indicating the fastness of different pigments and their incompatibilities when mixed. In a special section at the end of this essay we will provide an up-to-date list of all the materials used in paints and varnishes, with all the indications of their properties.

The fourth series of causes of alteration of paint concerns the very method of painting and the way in which paint is used. Artists assign too little weight to these considerations. I have met artists considered great who by no means concerned themselves with incompatibilities in painting methods: they alternated oils, gouaches and even pastels in a single work, without concerning themselves with their paintings' future destiny. In addition to this, an incurable fear of gloss causes them to thin oil paint so much that it loses all its consistency. Painters accumulate thick layers of tempera or gouache on the canvas, when these techniques by nature demand repeated thin layers. And so on and so forth. Every method of painting has its own prerogatives which we ought to take advantage of, but also its limits, which we must not attempt to overcome. There is no single painting method that can be adapted to all techniques and all caprices, and this may be a good thing, if it is true that the need to stay within material limits foments mastery of the art by requiring us to overcome difficulties.

To get back on topic, we must pay careful attention to how different paints are made. If a glue binding a faster pigment to a gouache cracks, peels off or fails to attach to its base, the painting is destined for certain death. If a drying oil contains excess metals, it will turn black, or if it loses its transparency for other reasons, the harmony of hue will be altered and the painting ruined. And so on. Let us be guided by our initial classification and seek to discover the different causes of deterioration and alterability in different kinds of paint.

The stability of paints of the first type depends on the physical form *(c.7r.)* of the coloured powder and on the nature of the substrate supporting it. The cohesive power of certain materials, such as kaolin, for example, is very considerable. Certain silicates, certain hydrated oxides, have a cohesive power so strong that considerable force is required to take them apart. Anyone may see for himself how natural Siena earth, which is in part iron silicate, becomes rock-hard when mixed with water. The same is true of hydrated chrome oxide green (emerald green) and many other colours. Others have a repulsive power, such as certain blacks. These phenomena appear to be due to differences in electrochemical tension, which may be positive or negative. This is commonly referred to as attraction or repulsion. In-depth study of these properties might open the path to understanding of many of the reasons why paintings crack. A good study of this topic is that by <Moreau Vauthier>, easily available in Ugo Ojetti's excellent translation, to which the reader is referred.

To provide painters with assurance of the stability of these paints, we might merely note that the oldest paintings known to us, which some archaeologists date back to... 50,000 years ago, are painted with a sort of fresco technique, and they are in very good condition! And so ultra-fine powders and bases rough enough to capture the pastel are the essential conditions that have allowed these paintings to last so long. To prevent mould formation, which is very likely especially in blacks, as they provide the perfect conditions for growth of these microorganisms, it is advisable to spray paintings with formaldehyde solution, which is perfectly innocuous. To dry any moulds that may have appeared before they corrode the paint, before applying the formaldehyde, spray the painting with pure alcohol and leave it in the sun. Press gently with the tip of a compact ball of cotton wall on the mould halo and it will disappear. If the roots of the mould have penetrated deep into the canvas or the mould comes from deep within the layer of paint, its deterioration is irreversible, and all we can do is touch up the painting.

Can pastel be fixed with a fixative? The answer is definitely no, and the reason for this may easily be explained. The surface of a pastel painting is an agglomerate of coloured particles which are not juxtaposed but each surrounded by an empty space,

13 Charles ***Moreau Vauthier****:* La peinture *(Paris 1912-1913), translated into Italian by Ugo Ojetti (Bergamo 1913).*

(c.8r.) however infinitesimally small. They therefore offer a surface area for reflection of light which is much greater than the area painted. The result is that the amount of light reflected is proportionately greater than that in a simple painting with a perfectly united surface, that is, a glossy surface. If any kind of agglutinant is used to fill in the gaps between the particles of colour, totally or partially, the result will of course be darkening of the hues, naturally accompanied by a brightening of their intensity. And because, as we will see in our specific analysis of coloured pigments, the index of variation in hue between dry and wet varies greatly from pigment to pigment, the unbalance in hue of the fixed pastel is inevitable.

(c.gr.) We must therefore go back to the classification
made earlier to identify the causes
of alteration of paint in the different
techniques appearing in the list.
The stability of paints in the first series
depends primarily and almost exclusively on the nature of the
substrate. A base that is too smooth or can itself deteriorate
will cause the powder adhering to it to be detached.
But we must consider that the adhesive and cohesive power
of many substances is higher than it seems.
The oldest paintings we know of, the cave
paintings ... are simple marks made without
glutens and they date back thousands and thousands of
years. The greatest hazard when it comes to pastels is the
greater possibility of alteration in the light
due to the lack of protection enclosing the
pigment. And old pastels are in fact almost always
whitened and chalky.
There are several causes of alterability of paints in the
second category. Discolouring of watercolours may
also be due to lack of protection of the gum,
which may deteriorate producing moulds,
blackening and cracking. The disintegration of
paper glue can also weaken this kind of paint,
which is on the other hand very solidly anchored
to the paper. Lean gouaches containing glues
are weaker. Glues are highly alterable
materials, both physically and chemically,
and offer an excellent bacterial culture,
causing moulds and irremovable
stains.

(c.9v.) The greatest danger is caused by crumbling of colour due to alteration of the glue, which no longer acts as a pigment binder.

The difficulty of restoring paintings of this kind ought to suggest to artists that it is best to use top quality materials. It is a well-known fact that many gouache formulas include agglutinating ingredients which give the paint a false velvety look at the expense of fastness.

Encausto painting requires lengthy discussion. This name is applied to a large quantity of concoctions. The basic ingredient in this kind of paint is wax, which is added molten or semi-molten to terpenic solvents or hydrocarbons, or partly saponified with soda or potash. The addition of oils and resins makes this kind of paint unstable and chaotic. But the greatest defect of this kind of paint is the tendency of wax to melt and rise to the surface of the colour. This is the operation the ancients performed by heating the finished painting. It is clear that this physical instability of the wax will cause shifting and alteration, especially in the oxidation of the oils added to it, with deleterious consequences.

Coloured pencils are pastes mixed with wax in dry form, and are therefore quite stable, provided they are made with quality materials.

Varnish paints are only as inalterable as the varnishes from which they are made. They are then used as the basis of all (c.10r.) further preparation and solidification.

They are therefore valuable as an intermediate product, saving the artist the tiresome task of mixing and remaining inalterable for an indefinite time in their tubes (See Chapter 3).

And now we come to the prince of all paints: oil paint.

The history of painting from the 16th century on demonstrates that the technical possibilities of painting were multiplied a hundredfold with the adoption of the techniques discovered by Van Eyck and Antonello.

And the great damage suffered by many of these paintings demonstrates that these techniques have not always been applied correctly. But to condemn them in a snobbish attitude originating in the late nineteenth century and still commonplace today is an exaggeration.

The fact is that this kind of painting is still the easiest, the most practical, the fastest and most inalterable, if used as it ought to be.

First of all, we need to expose a myth.

Oil does not alter paint and it is not the main reason for the truly impressive state of ruination of many of these paintings. On the other hand, oil is the most preservative form of colour, and with the exception of a slight yellowing of linseed and walnut oils, it has no malignity at all.

It is the concoctions that are added to them that alter their valuable qualities.

It is in fact obvious that a medium composed of two or more liquids with a different tension,

*11 Jan **Van Eyck**, **Antonello** da Messina: painters considered in anecdotal art history to be the inventor and the promoter in Italy of oil painting, respectively. The first worked in Flanders (1390-1441), while the second (c. 1430-1479), according to Vasari, travelled to Flanders and brought back the new painting technique. Modern critics consider it unlikely that the two ever met, taking into account their dates.*

(c.10v.) one of which (the oil) is still in a state of stabilisation, that is, of decreasing allotropism, will definitely produce stabilisation shocks similar to those following an earthquake and with the same consequences, if the other elements do not follow the oil's internal shifts as it oxidises with enough elasticity.

If we add to this alteration of the substrate and general alteration of dilatation and shrinkage due to atmospheric causes (heat and moisture), it is easy to explain the physical alterations of paints, which range from fogging ("chancy") to peeling, cracks, blisters, flaking and therefore to disaggregation and pulverisation of the paint.

The painter's primary concern should therefore be making sure that oil paints are really oil paints and above all that they do not contain any wax.

To check for the presence of this poison in the paint without performing a chemical analysis, all we need do is heat a little bit of paint to about 80° C. The wax will melt at this temperature and rise to the surface.

When the paint has cooled and dried, if we rub the matt surface, it will become shiny.

Another serious problem is the presence of excess resins either dissolved in the oil (oily varnishes) or in other excipients (terpenes or hydrocarbons).

Here the opposite happens. As they dry, the resins settle at the bottom of the coat of paint,

2 ***allotropism****: what we mean here is the transition of a composite from a phase of greater volume to one of lesser volume, with reference to the change in size between the mixture when just applied and how it settles when it polymerises.*

12 ***chancy****: in the sense of uncertain, indistinct.*

(c.11r.) so that the misleading shine on the fresh paint quickly disappears during the concomitant process of oxidation and drying of the paint.

In oil paints made with colour and pigment alone, on the other hand, resinification from above adds splendour and brightness to the hues with time, and appropriate glazing with varnish when the paint is dry will ensure that the colour produces the best possible yield.

Other causes of deterioration may be found in the quality of the pigments and the fastness of the substrate.

We shall not list pigments according to lightfastness and resistance to natural agents here.

At the end of this treatise we shall provide a reasonable list of the fastest pigments and their quality. On the other hand, essential to discuss substrates here, in view of their considerable importance. That canvas is the ideal substrate in many cases is proven by its durability and convenience.

But the problems lie in choice and preparation of the canvas. It is hard to paint directly on the fabric, especially when producing a finely detailed painting. And canvas is not only easily torn but presents other problems because it absorbs moisture. So while linen and hemp shrink with dampness, cotton and, worse yet, synthetic yarns tend to stretch.

Tests have been performed with "medieval" fabrics in which linen yarn is used to compensate cotton, but this creates highly damaging diagonal tensions. The best canvas is certainly *(c.11v.)* that made of hemp, which is highly resistant to forces and stays relatively the same if pulled over a drum when dry over a solid frame that holds it still. As for preparations such as oil-based ones, they can easily "cook" the fabric, while glue preparations make the canvas fragile. The best preparations are those of significant thickness, normally prepared with gouache or chalk. Direct oil preparation often flakes off, especially on linen. Early nineteenth-century canvases are extremely fragile and transporting the paintings is often a very hazardous and delicate operation. Preparations on wooden boards are certainly optimal provided the boards have been stabilised well and prevented from warping and distortion, provided woodworm is prevented.

And so, when the elementary precautions described above are applied, no kind of paint presents more pros or less cons.

Maximum yield and maximum intensity of hue can be obtained from oil paint using all techniques, from the thickest paste to the lightest glaze. This type of paint also offers the huge advantage over all the others of colour which stays the same as it is when applied to the canvas, without the great changes characteristic of gouaches and watercolours as they dry, a difference which is most striking in the fresco technique. In this regard, it is important to dispel a myth regarding the apparent (*c.12r.*) change in hue which sometimes appears to the artist. It is a well-known fact that, due to a known optical phenomenon, a coloured pigment in the powder state acquires greater intensity of hue when its many reflective surfaces are drowned in a uniform transparent medium such as oil or varnish. But in the presence of an excess amount of a medium which has a different refraction index from the pigments, there will often be reflective distortion and the colour will lighten a little more. This is the phenomenon which occurs in glazes with very runny paint, which gives an impression that the pigment base has grown. This phenomenon is familiar to gouache painters, for gouache colour tempered with a little white darkens more strongly as it dries. The same phenomenon occurs in glazing with varnishes containing only a little colour. But it is entirely an optical phenomenon. For example, the behaviour of certain silica-based paints of a colloidal nature such as authentic natural Siena earths is renowned: they appear much lighter than their true hue as soon as they come out of the tube. The true colour appears a few hours later. So what is the origin of this hostility, sometimes truly morbid, towards oil painting? The primary source of this prejudice is the great nineteenth-century restorers led by Secco Suardo, considered a master of technique by many other prominent restorers. The special requirements of restoration technique are incompatible with even minimal variations in hue.

(c.12v.) Now all oil painting becomes slightly more subdued in hue with time, however evenly, and this is unacceptable for someone who has very patiently managed to obtain the exact copy of a colour in an old painting. And application of a living material such as oil to a fossilised skeleton such as an old painting will, due to the resulting oxidation, result in certain changes to the long-established balance of the old painting. And so it is only right that wise restorers should ban use of oil paints in restoration work. But there is no reason in the world why this ban should be extended to new paintings, and all the ills of painting be attributed to this precious and irreplaceable friend of the painter. Good oil will not rot and will with time transform into the hardest and most splendid gem of amber. Good oil is ductile and obedient, and will adapt to all caprices and demands. But the ancients did not use oils, they used gouaches, and they are still fresh and healthy years later, brighter and clearer than any oil painting. That's the way it is. But what living painter has the pointillistic patience of Vivarini or the early Bellini, to work to obtain the stability that can only be obtained with an enormously absorbent base? And what will become of the wide, all-embracing great brushstrokes

*27 **Vivarini**: most likely intended to refer to Alvise Vivarini (Venice, c. 1446-1503/5), an artist profoundly influenced by Giovanni Bellini and Antonello da Messina; **Bellini**: most likely Giovanni Bellini, also known as Giambellino (Venice, c. 1433-1516).*

(c.13r.) of Tintoretto or Caravaggio? But the oil that we are proposing banning (only to replace it with what? wax?) was always a part of the old gouaches, or at least the strongest ones. Egg, which was the most commonly used gluten, contains 20% fat, a drying oil of its own which is much less pure and much darker than linseed oil, for example. And it is precisely the oil in egg that gives gouaches this inalterable quality that so many enemies of oil admire. And here we come to that miraculous and not entirely understood process of electrochemical cohesion known as emulsion. The presence in egg of a powerful and complex emulsifying substance permitted dispersion of oil in watery gouaches, promoting even, stabilised dispersion.

THE FATTY GOUACHE OF THE ANCIENTS

As we have seen, the gouaches of Theophilus and Cennini contained egg, either whole with both yolk and albumen, or with yolk only, with the addition of latex fig or other agglutinating substances. Latex is itself an emulsion, like a number of natural products containing fatty substances dispersed in water: typically milk, egg and the lymph of many plants. Fig latex is a substance similar to rubber, to the secretions of other euphorbiaceae of *(c.13v.)* many members of the lettuce family and certain resins mixed with gum which, soluble primarily in water, become insoluble with evaporation due to a particular chemical transformation. To these basic ingredients the ancients added resins which partly saponified the fat in the egg and facilitated dispersion in the emulsion. The emulsifying power of egg yolk stabilised the emulsion quite well. But the preparation was quite likely to rot due to the presence of many organic additions, requiring addition of a certain quantity of vinegar, which also served to precipitate the albumen and make it insoluble. This preparation had to be made at the time of use, so that the colour first ground with pure water had to be added at the time of use to the *tempera,* that is, the emulsified mixture. This was a laborious and difficult technique which only the patience and sobriety of the old masters of the art could manage.

Technicians at all times in history often attempted to manufacture these preparations industrially, but their success rate was very low, so that, while they used other types of emulsions, particularly containing casein ammonia and, more recently, glycerophtalic resins, egg gouache was once more relegated to the studies of certain (c.14r.) patient makers of paint who...

Our technicians have also been addressing this intriguing problem for some time. Gouache can achieve effects that cannot be obtained by any other painting method in terms of freshness, sumptuousness and clarity of hue. Its lack of viscosity makes it valuable for certain clear, nervous pictorial accents which the dry pastiness and brilliant consistency of oil do not permit. Even degreased oil cannot give that special, sincere material nature characteristic of gouache. It is well worth dedicating further study to the matter.

But our technicians quickly gave up the attempt to stabilise and preserve paint made with egg and also neglected all other irrational manipulations involving use of albumen, resin soaps, casein and more. These are unstable raw materials which yield only poor aesthetic results.

They turned their attention instead to the natural process of egg. They studied its essential components in view of their adhesive and emulsifying power and became convinced that it would be possible to make the preparation synthetically by a similar procedure, scrupulously observing the proportions of the components of egg involved in the emulsions. Basically, what they needed to do was first of all eliminate all those accessory substances in egg that had other *(c.14v.)* biological functions not essential for our purpose, with the benefit of doing away with many of the causes of deterioration of the tempering mixture. And so all the sulphuric parts were rejected, along with many ingredients of value for other applications which slow down the chemical process of interest to us. Of the remaining elements that were found to be indispensable to and characteristic of the process of emulsion, we attempted to use the synthetically pure product, or at least to replace it with another very similar element. And so we came up with a synthetic reconstruction of the part constituting the egg and essential to it. Drawing on the experience of the practices of many of our predecessors, we then added, with all the due precautions, those resins which it seems the ancients also added, and which have given definite, decisive evidence of beneficial effects in modern gouaches. As the final, very useful result, we were able to eliminate both useless vinegar and other disinfectants, made redundant by the perfectly aseptic formula we proclaimed. A final but important concern was product shelf life in the tube. Here too, an ingenious solution to the problem allowed us to prepare an emulsion which, due to its own internal constitution, could not be broken up and which became even more

(c.15r.) intimate and long-lasting with time. Everyone knows that emulsions are colloidal mixtures in which a normally insoluble material is dissolved in another material, simulating a solution, the emulsifying stability of which will be proportionate to the degree to which the product is dispersed. When the balance thus obtained is broken, the parts separate and the emulsion is broken. Emulsions of, for instance, oil in water are normally classified in the two formulas oil in water, if the proportion of oil is lower than that of water, which acts as a solvent; or water in oil, if it is oil that incorporates a smaller quantity of water. The same applies to true solutions in both categories. It is difficult to go from one class to the other, which is why the emulsion breaks up at a certain point. This breaking point will of course depend on the extent of homogenisation and the nature of the solutions.

To get round these problems, which may arise in the emulsion itself with time, we managed to create, so to speak, emulsions of water in oil which may potentially be transformed into emulsions of oil in water. And so our gouaches are preserved in their tubes like ordinary oil paints. Not only this, but they may, if you wish, be thinned with solvents other than *(c.15v.)* water.

These brilliant results, which represent an industrial solution to a problem which has been addressed repeatedly but never entirely solved, do not represent achievement of our goal, but merely a step on the way.

Stopping a moment to enjoy the product we offer artists today, we do not consider our purpose achieved, but as always, we ask users of our paints to help us to perfect the already conspicuous quality of this product.

Characteristics of silver label Maimeri fatty gouache.

Maimeri fatty gouache is water-soluble and requires addition of water, from a few drops to a maximum thinning the paint to the consistency of watercolour.

Of course the more it is thinned, the less regular the dispersion of pigment will be. It does however produce beautiful, even, dense surfaces.

The hardness of the water, that is, the presence of minerals (basic?), will hinder dispersion and tend to break up the emulsion. We have therefore prepared an emulsifying medium which may be added to the paint when the balance appears to begin to break up.

Maimeri fatty gouache is subjected to a regular oxidation process which lasts for anything from a few hours to a few days, depending on the substrate to which it is applied, after which it gradually becomes insoluble in water.

(c.16r.) Different colours have different solubilisation times depending on the catalytic power of the pigments used. And so the painter mustn't be in a hurry to varnish gouache paintings, as their insolubility is definitely lower than that of the stronger solvents, such as spirits and petroleum. We supply an excellent alcoholic varnish which can be sprayed on quite early without damaging the painting. But we must not forget that gouache is a technique for the patient, and we must therefore consider the time coefficient a precious ally.

The dispersive power of gouache is huge. To convince ourselves of this, all we need do is touch a wet surface with a tiny bit of paint; the inadration of the colour will be lightning fast. This is a valuable property for obtaining uniform, evocative hues.

But this kind of paint may effectively be used both in paste and in discrete broad brushstrokes with only a little water. Ordinary water is always a necessary aid for obtaining the clear background surfaces characteristic of gouache painting.

We have already discussed the phenomenon caused by excess water. This may be attenuated using other solvents. Excellent results may be achieved with vegetable or mineral spirit. Dispersion is clearly less and paint made in this way

17 ***inadration****: probably meaning "dispersion".*

(c.16v.) is similar to oil paint, even while maintaining the properties of gouache. This does not mean that different solvents can be mixed up and used alternatively in the same paint. Regularity and order in painting proceedings are the primary basis of stability. But it is certain that with paints of this kind, it is possible to layer one over the other (provided they are completely dry) layers of water-based gouaches, layers of essence gouaches and even layers of oil paint, without the painting as a whole suffering as a result. What we must always avoid is painting on surfaces which are non-absorbent or shiny, such as oily canvases, the shiny side of masonite, glass, ceramics, and polished and burnished metals. The best substrate is still definitely the classic preparation of seven layers of glue and plaster. Paper is excellent, for it admirably embraces paint. Admirable, suggestive effects may also be obtained on the surfaces of matt paper for pastels or on wool, fustian or other compact, absorbent fabrics.

Varnish paints for restoration work

Our series of paints for the fine arts was still missing a product of great valuc to artists.

(c.17r.) We filled in this gap with this precious new product.

The presence of oil in paints is, as we have seen, often intolerable. This is the case of paints for restoration, a type of work which demands paint that absolutely will not move, in the sense that it will change in neither hue nor gradation with time. Gouache meets this need perfectly, but while it does not in itself change hue, it looks different when dry or wet, and final varnishing does not normally restore it to its exact appearance when wet. This is obviously a result of the fact that the solvent in the varnish has a different index from water, and there is as yet no stable water-based varnish.

Varnish paints meet this need, for their solvent is the same as the solvent used in the final varnish, and so there can be no variation in hue between the original wet paint and the final varnished paint. Not only this, but these paints withstand addition of resinous varnish or any other ingredient the painter should wish to add. They thus form the basis of an infinite number of preparations which the artist was in the past obliged to create meticulously from scratch, by rudimentary and imperfect means, without ever being able to obtain perfect homogenisation or even a tolerable grind.

Our varnish paints for restoration work

(c.17v.) have the precious quality of being perfectly anodyne. They contain a very tiny amount of resin, no more than is necessary to hold them together. The solvent is one of the most volatile and inert, and leaves no trace of itself once it evaporates. And so, the painter seeking a personal formula may add to it as many ingredients as he likes, confident that the colour will be perfectly ground; and the decorator may add his formulas to make his precious enamels, mastic, etc.; the restorer may, by skilfully dosing his precious varnish, obtain the closest imitations of the ancient blends; the painter on fabrics may dilute the paint dried in his starches; and the encausto painter may incorporate the waxes and essences of his concoctions in it. And our list of examples might go on indefinitely.

Many artifices believe they can replace grinding a gossamer-fine powder in mills with powerful machinery by hand grinding, as if grinding were merely a matter of patience and time. Nothing could be more wrong. The task of grinding is not so much one of crushing the pigment as of homogenising it. All substances have a particular adhesion index which ensures that they are arranged in a particular order *(c.18r.)* around the other elements with which they are mixed. Some elements have an affinity, others a repulsion. These are the electro-chemical phenomena underlying the study of mixing solutions in emulsions. The purpose of grinding is to obtain the greatest possible homogenisation, consisting of the greatest regularity and symmetry of the function of different compounds in the mixture. This result cannot be obtained without expending a vast amount of energy. It is impossible to seek to obtain this result without adequate equipment. Anyone may try this by mixing or grinding a little zinc white with oil: the quantity of oil required to obtain a reasonably fluid paste will be much greater than the amount required in good mechanical grinding. Not only this, but the amount of time it takes to rub the white by hand will cause it to turn grey, while in the mechanical process rubbing contact is minimised. This is because zinc white has a very low cohesion index. To the contrary, Sienna earth ground by hand will never absorb as much oil as it should by nature, because it is impossible to achieve the pressure required to saturate it and so it will be impossible to obtain the transparency that characterises this material. These are only two limited examples, but it is impossible to obtain true results without adequate equipment.

(c.18v.) Black label varnish paints for restoration work are therefore a valuable aid for painters suitable for all kinds of uses.

On varnishes and thinners
(...)

A LIST OF THE COLOURS MOST COMMONLY USED IN THE MODERN COLOUR PALETTE.

White colours

Silver White, Kremniz White, White Lead, etc.: a basic lead carbonate, with strong covering power, a good matt white, lightfast but perishable when lead is reduced to metal due to some kind of reaction, especially upon contact with sulphurous emanations, obscuring or yellowing the white hue. Its greatest virtue, apart from its covering power, which has been surpassed by other substances, lies in its suitability for saponifying oils, producing a very hard mixture with time which is extremely fast and long-lasting. Very poisonous, and must be used with great care.

5 Title of the chapter later developed in ms. at lines 25r.-28v.

(*c.19r.*) *Zinc White:* Zinc Oxide obtained by sublimation of the mineral. Ultra-white, almost totally inert in the usual blends, it is non-toxic and is now the most commonly used white. May be mixed with any colour, and has the sole defect of less coverage and a tendency to separate from the oily medium if not sufficiently homogenised, producing characteristic cracks (parallel and ?) see). It is therefore not recommended in watercolours due to its tendency to hydrate and lose many of its qualities.

Titanium White: Titanium Sesquioxide. A relatively new colour which is rapidly gaining ground in the industry, thanks to its admirable qualities of coverage, inalterability and ... Difficult to homogenise on its own, in weighted mixtures with other whites (zinc, barium, etc.) it produces marvellous blends which are much appreciated by artists.

Lithopone ...: Used only in watercolours, in place of zinc white, and in the preparation of base coats. Not recommended for use in paint due to its tendency to lower in tone (turning grey) in the light, though it turns white again in the dark. We must not forget that the presence of sulphides in its composition makes it entirely inappropriate for use in contact with lead paints.

(*Egyptian Blue* ? ($CaCu\ Si^{4}O^{10}$) copper and lime silicate).

(c.19v.) *Notes on Oil* (from Guareschi): Heraclius spoke of oil painting in the 10th or 11th century: De omnibus coloribus cum oleo stemperatis.

Naples Yellow: lead antimony. Often contains lead oxide: must not be touched with iron, so no yellow earth and no Prussian blue.

German Blue: basic copper carbonate $2Cu\ Co^3\ Cu\ (OH)^2$ is transformed into malachite $= CuCo^3$ $Cu\ (OH)^2$ copper acid carbonate

Mountain blue: cobalt or enamel silicate

Verzino lake = Brazilian wood

Lapis amatite = sesquisoxide red hematite of iron Fe^2O^3

2 ***De omnibus coloribus cum oleo stemperatis****: citation from Pseudo-Heraclius, that is, an aggregation of other medieval texts with Heraclius'* De coloribus et artibus romanorum, *present only in the lat. ms. 6741 of the National Library in Paris, also known as* The third book of Heraclius. *The manuscript in question was the basis for the Merrifield and Ilg editions, resulting in attribution of a a series of other texts to Heraclius, including: a paraphrase of a number of chapters of Heraclius regarding glazing; a long quote from Isidoro of Seville; various recipes probably originally from northern France; the interpolated version by Faventino (cf. S. Baroni, "De generibus colorum et de colorum commixtione. Faventino interpolated", in* Quaderni dell'Abbazia, *Fondazione Abbatia Sancte Marie de Morimundo and Museo dell'Abbazia di Morimondo, year XV, n.u., 2008, pages 55-66); tables for mixing colours in the so-called "DCM family". The passage in subject belongs to a section on painting on stone: "Tunc vero poteris desuper de omnibus coloribus cum oleo destemperatis pingere" (in truth you will only be able to paint over it with all colours dissolved in oil). This passage was amply commented on and debated in the context of nineteenth-century speculation about the origins of oil painting.*

14 *At this point the manuscript shows a draft of the preface, which the author deleted with a series of lines forming a St. Andrew's cross. The rejected text, which we present here to permit comparison with the final version, in a note at the beginning of the treatise - its logical location - runs down to line 20v. (r. 18).*

(*c.20v.*) *Blue colours*

Cobalt blue: Cobalt aluminate. A very fast colour offering good transparency and good yield. May be mixed with any colour. In artificial light it turns a dirty purplish hue. Dries quickly.

Cobalt light blue: Cobalt stannato. A very fast colour offering fair coverage. Like many colours tending toward greenish, acquires added splendour with white. Has the property of staying blue under artificial light. May be mixed with any colour, dries out less than aluminate.

Ultramarine: Sulpho-silicate of aluminium and of soda, of indefinite chemical formulation.

(*c.21r.*) A very bright colour which is resistant to light and natural agents.

Offers good coverage, but darkens in artificial light, so it is recommended when preparing skies that they be brightened up with a touch of emerald green. Is not resistant to acids. Suffers what is known as Ultramarine malady, consisting of a tendency to separate from the medium surrounding it, causing drying. Slow to dry. Medium oil hold.

Prussian Blue: Basic ferrocyanide.

A very intense lacquered colour, with great colouring power but poor lightfastness.

Even in compositions, it acts as a reducer and alters colours. Not resistant to alkalis, it is also poisonous. High oil hold.

Cadmium Yellow: Cadmium Sulphides.

Available in a very wide range of hues, from lemon yellow to deep orange. Lightfast, inert with very good colouring power and coverage, these yellows entirely replace chrome yellows.

The lighter shades are slightly less stable than the medium and dark ones and lose a little of their gloss though without changing hue. They are incompatible with lead paints only due to the chemical action of sulphides. Rather slow to dry. Low oil hold.

Chrome yellow: Lead chromates ranging

(c.21v.) from acid chromates to definitely basic chromates, from light yellow to orange. Colours with great covering power but poor resistance to light and natural agents. They blacken quickly in the presence of sulphurous fumes. With little oil hold, they are not resistant to alkalis and are not used in fresco painting, except for the red, which is solid. Rapid drying.

Naples Yellow: Lead antimony

Antimony Yellow: lightfast colour. Heavy, with good coverage, highly resistant to lime. In paint it has been replaced by...

Zinc Yellow: Zinc chromates and potash. A good lemon yellow, very lightfast, with poor covering power, medium weight and medium drying time. Due to the solubility of the potash bichromate in its composition, it cannot be used in watercolours.

Ochres and coloured earths

Coloured earths were definitely the first colours to be used in painting. Men were easily able to note their fastness in the natural state and chose among the many hues offered by nature. Coloured earths are normally clays (Aluminium silicates) coloured by the presence of iron oxides or hydrates, by either mixing or absorption. Chemically speaking, they are true iron lakes, in which the iron oxide is fixed in alumina salts.

(c.22r.) The base often consists of diatoms or fossil powders, the structure of which gives them great absorbency, to the extent that they can absorb very high percentages of colorants. Earths from Vicenza and Verona, which provide the base for green earths, absorb organic greens in very high quantities.

Italy is very rich in valuable coloured earths, from Verona greens and yellows to those of the Brescia and Bergamo areas, to the precious earths of Mt. Amiata near Siena and those of the islands of Giglio, Sardinia and Pozzuoli.

Yellow ochres are composed of iron hydrate or limonite. Natural oxidation transforms them into red ochres or amatite, <Cennini's> famous lapis amatito, which many erroneously confused with amythest.

The pinkish natural earths of Pozzuoli, the red earths of Malaga with their very high iron oxide content; the base for English reds are Indian <ones>, etc. They may arbitrarily reduced to red with firing, creating burnt Sienna. The earths sometimes contain manganese, forming Umbers, and sometimes contain tar, as in Cassel earths.

A separate class is that consisting of synthetic earth oxides, which range from light yellow to purplish brown and red. These are very fast colours with very high covering power.

17 ***Cennini's famous lapis amatito****: Cennino Cennini,* Il libro dell'arte, *chap. XLII. On the nature of a red known as amatisto or amatito. This controversial passage has generated different interpretations of the nature of the mineral described by Cennini. On the issue, cf. S. Baroni, "Riducendoti al triare de' colori. Prescrizioni sui colori nel Libro dell'arte di Cennino Cennini", in* Acme, *vol. LI, folder 1, 1998, p. 53-64 (p. 59-60).*

(c.22v.) Some comments on the earths: their oil hold ranges from maximum in the case of natural Sienna earth and green earth, natural Sienna, umber, yellow and red ochres and iron oxides. Drying time is average for all, and the umbers dry quickly due to the presence of manganese. Lightfastness is total, resistance to natural agents absolute. Sienna earth has been accused of darkening in light. This is wrong. The phenomenon by which colour just squeezed out of the tube onto the palette (darkens) is a optical rather than chemical, and is due to the excess liquid it contains.

Cinnabar = Vermilion. Mercury sulphide. A colour with strong coverage, very heavy, with very poor oil hold. A very bright colour. The more orangey hues are the brightest and purest. Tempered with white, it maintains its hue quite well. But the darker hues tend to be soiled with white. Lightfastness is good if used alone. Blackens in the presence of lead paint or sulphurous fumes.

Cadmium Red: Cadmium sulphoselenide. Much lighter than cinnabar, which is not as opaque but still has a very strong colouring power. Ranges from an orangey hue to purple, losing purity of hue as it darkens.

The most purplish hues are violet and dirty when mixed with white. Cadmium orange (c.23r.) has the curious quality of changing to red when tempered with white, while most colours when mixed with white tend toward the centre of the spectrum, that is, towards a yellowing green.

Lakes of all kinds

Lakes are colours produced by fixing a soluble organic colour on an inert white mineral base, or better yet, a transparent base, which gives it consistency and solidity. Bases are normally barium or calcium alumina salts, depending on the desired weight and transparency. Sometimes natural siliceous earths such as Bentonite are used, fossil powders, <...> earths, Verona earths, etc. The finest lakes are those fixed with alumina. Lakes' lightfastness depends on the nature of the organic colour. The most stable lakes include natural madder lake with its synthetic (allotrope) ?, Alizarin, colours a beautiful deep purplish red ranging from cherry red to purple violet. Colours that are highly resistant to light and natural agents but not to lime, with good oil hold and few drying agents. Madder and alumina lakes, especially madder lakes, can be subject to the annoying and well-known phenomenon of gummy thickening (gelling). This is a complex phenomenon of internal oxidation (polymerisation) of the oil due to causes which are not well-known and often *(c.23v.)* facilitated by the presence of alumina, which seems to act as a catalyst, and the acidity of the oil.

Aniline colours and organic synthetic colours

At one time, speaking of aniline and organic colours was considered an unforgiveable sacrilege in artistic painting technique. Now, organic colour technique has reached the point that banning certain magnificent hues from the palette has become an outdated and unnecessary prejudice. A number of <...> reds, <...> blues, pigmented greens, certain admirable Wolframium lakes, certain <...> yellows are very fast and many of them are even resistant to lime and fresco painting. But they must be used with great care, especially in fresco painting, because the many different names under which they are sold are not always reliable. These are, in any case, definitely the colours of the future, with their ease of mixing, their intensity and hue.

Guignet Green or Emerald Green: Chrome oxyhydrate. A green offering great resistance to light, acids, alkalis and heat (up to 500°), very transparent, with fair colouring power. Good oil hold and good drying; may be mixed with any colour. Excellent for fresco painting.

Chrome Oxide Green or grass green: a olive green with great covering power, high weight and little oil hold. Has the particular quality of not changing tone from dry to wet, in either water or oil. Very resistant to lime, acids and light, perfect for frescoes.

(c.24r.) We discussed green earth in the section on natural earths. Other lightfast earths include Cobalt Greens, which are combinations of Zinc and Cobalt. A good cold green, they offer medium coverage and good oil hold, with good drying. They may be used in any mixture. Maimeri was the first to introduce, in addition to light green, an excellent dark green offering good coverage, useful for making up deep shades of green. Better in oil than in water, it resists the fresco technique very well. *Schweinfurt Green* or Paolo Veronese green: has a magnificent cold, light green hue. Very poisonous, with little covering power, quite lightfast, it is water-resistant and, when mixed with other colours, either increases or decreases their tone. It can very well be replaced by brighter surrogates with better fastness. Chrome greens, Zinc greens and Cadmium greens are made up of Prussian blue with <...> yellows and ash and zinc yellows, in the former case, and very lightfast, made of chrome hydrate and cadmium yellow like the nearby permanent

(c.25r.) green, in the latter case.

Ivory Black: made up of calcified purified bone black. A deep black with good covering power, good oil hold and slow drying.

Sepia: a natural product offering good fastness.

Grape Black: a vegetable black which may alter due to the presence of organic bitumen.

Cobalt Violet: Cobalt phosphate. A bright colour with fair coverage, but transparent, it is prepared in two hues, light and dark. Resistant to light and to natural agents. Has good oil hold and dries rapidly. Not useful in fresco painting.

Mineral Violet: Manganese Phosphate. May replace dark Cobalt Violet, similarly bright.

Excellent lightfastness, good oil hold and good drying.

Many of the other colours are either blends, or should be rejected as overly ephemeral.

OF THINNERS AND VARNISHES

Today's paint industry makes paints for artists packaged in tubes of an average consistency, ready

In the current configuration of the manuscript, the text proceeds from 24r. to 25r./v., to return to 24v. and continue on 26r. The correct sequence of the text is indicated by the numbering, missing the no. 3, in red ink at the top left. The pages involved are part of a two-page sheet of legal paper, and the disorder may easily be explained by the custom, typical when writing in ink, of working with the sheet of paper open, covering two pages with writing and then turning over the sheet of paper to cover the other two. The subsequent cutting of the paper generated the existing sequence.

(c.25v.) for use. And yet depending on the type of work to be done, the artist needs to have suitable thinners at hand to facilitate application and fluidity. In watercolour painting, water is the best solvent, for both it and the water in the mix are destined to disappear as they dry. But in oil painting, the amount of oil added is the amount determined by the pigment's so-called oil hold, so that addition of a greater quantity is without doubt redundant. And so it is necessary to add, as temporary thinners, volatile essences which will evaporate before the oil begins the process of oxidation. But these additions also facilitate settling of the oil-colour combination, dragging the oils down to the bottom toward the surfaces that absorb them, causing what is known as drying, which takes the form of dulling of the painted surface in the areas where thinner has been used, creating an obvious unbalance in the parts of the painting thinned. It is therefore a good rule to add to the volatile thinner a skilfully dosed part of oil with greater or lesser drying power, depending on the result desired. On the basis of this principle and as a result *(c.24v.)* of practical experience, Maimeri makes slow media, drying oils and painting oils. Solvent is a petroleum product which evaporates completely, leaving no residues. On essence of turpentine. This precious solvent with its very strong cleansing power also has significant power to aid oxidation. Unfortunately its activity as an product that oxidises may harm the painting, especially if its oxidation process has already begun. And so it resinifies and becomes sticky. Not only that, but, depending on its origin, it may or may not go black, or at least darken. This means it must be mixed thoroughly with fluid pastes and applied in thin layers, while if incorporated in big, thick pastes it may cause problems. Mineral essences do not present this hazard. Use of essence-based varnishes in mixing and thinning paints is reproachable. This warning applies above all to varnishes. In thick layers of colour, the imprisoned varnishes destroy the equilibrium of the harmonious

(c.26r.) oil-colour (duo). Applied as a glaze
or thin paste, it cannot cause damage if a varnish is used
which is capable of amalgamating well with the oil in the varnish.
Vibert may exaggerate when he demands that the resin
be soluble in the oil in the paint,

while the opposite is always true,
that is, that oil is soluble in varnish.
The important thing is, as Vibert says,
that there be an affinity between the varnishes and the oil,
which therefore excludes shellac, as it is not
soluble in the presence of oil. Similarly, waxes act as
insulation and, while they may profitably be used as
a final protective coat in encausto and mural painting,
no-one may add them to oil paint while working,
as the different layers would be separated
and would flake off like fish scales with time.
In conclusion, resins may
very well be included
in the composition of colours, but
in this case...

(*c.26v.*) For these too we will provide a detailed analysis.

Linseed oil: the oil best suited to making paints. Its inalterable fluidity and its rapid drying make it valuable. Its only defect is a tendency to yellow during the oxidation process, especially if the reaction takes place in the dark. It is therefore a good idea to expose paintings to plenty of light and even sunshine during pauses in the work.

Walnut oil: yellows less and is softer than linseed oil but can often deteriorate in the presence of water or even dampness, resulting in powdering phenomena. Much less drying.

Poppyseed oil: very white when properly prepared and absolutely inalterable, it will not yellow in light or darkness. It is however much less drying and must be corrected by suitable measures.

Turpentine essence: already mentioned in the previous paragraph. Let us add that the best come from France and Spain. Evaporation point of around 150°.

(c.27r.) *Petroleum essence*, Petroleum ether, Gasoline Petrol, White Spirit: these are all names for fractions of crude or refined petroleum.

Unlike turpentine essence, which has a single evaporation point throughout its entire mass, petroleum products may be broken down, so that solvents may be classified with two evaporation points, initial and terminal, and are graded on a percentage scale.

The most appreciated qualities are those which break down the most. Today, refineries present products of perfect purity and gradation.

They generally have lesser solvent power on resins, but offer the benefit of evaporating entirely without leaving any residues.

Lavender essence: a terpene like turpentine essence, but with a much higher evaporation point. Always leaves oily residues which resinify and turn black.

(c.27v.) VARNISHES

Varnishes serve to protect the surfaces of paintings and brighten up their hues.

They must therefore be inert, hard enough not to be scratched, transparent and as colourless as possible.

The solvent must evaporate entirely, must not corrode coloured surfaces and must be chemically inert. We must therefore first exclude varnishes made with powerful solvents, which would ruin the colours and could be hard to remove.

Oily varnishes, that is, those made with molten hard resins dissolved in drying oil with the addition of essences, cannot be used in the fine arts. These include Coppali, Flattings, varnishes referred to as Casse, etc. In addition to definite yellowing, they become insoluble with time in any solvent and therefore can no longer be removed.

Alcohol-based varnishes must be used very carefully and only when familiar with their ingredients, or when guaranteed by a trustworthy manufacturer.

The presence of shellac is very damaging

(c.28r.) to oil paint, because this resin does not amalgamate with oil. On the other hand, it is indispensable for the first coat of varnish on glue-based or fatty gouaches. The alcohol does not damage the tempera mixes and the resin can be added to fill in and surround the raw colour of these paints.

Maimeri produces special alcoholic varnishes for fatty gouaches.

Essence varnishes may be divided into turpentine varnishes and petroleum varnishes. The former are the well-known Dammar and Mastic varnishes.

Dammar varnish is a valuable varnish, perhaps a bit soft but very clear. In the usual preparations it often appears opalescent to some degree, in some cases entirely milky.

Due to a fortunate combination of, circumstances, Maimeri has discovered a way of preparing it in a perfectly clear solution.

Mastic varnish is one of the best varnishes when not adulterated with the addition of inferior resins. With good hardness, very glossy and pasty, its only defect is that it yellows slightly with time, especially if left in the liquid state.

As for petroleum varnishes made with the resin referred to as *(c.28v.)* normal, the reader is referred to Vibert's manual, a true authority in many things, despite the fact that he takes certain of his obsessions a little too far.

Maimeri uses this method to produce very bright varnishes using a formula by Dr. Carlo Maimeri.

These varnishes, which are used to make the new black label varnish paints for restoration, are really inalterable inert matter which will not yellow.

As they are made with solvents scrupulously chosen in the appropriate stage of evaporation, it is not advisable to thin them with different solvents. A different fraction of solvent may cause a part of the dissolved resin to precipitate due to saturation, clouding the solution either temporarily or by forming an insoluble deposit.

SUBSTRATES

Many problems with paintings may be attributed to substrates which are not suitable for painting. Apart from the substrate's own resistance, it may occur that the paint does not adhere well to the surface or that the preparation applied to the base, mix or <primer>, is not well attached to the canvas, cardboard or wooden board, *(c.29r.)* so that the paint detaches from the base with time. This happens with surfaces that are not absorbent enough or ...

These precautions are indispensable when painting with fatty gouaches. A substrate that is overly absorbent will thin the emulsion too much, so that it cannot solidify. A substrate that is not very absorbent will not attach sufficiently to the layer of paint, which even while solidifying on the outside, will stay weak in the innermost layers and will not resist handling during touching up and varnishing.

During the lengthy studies that went into the creation of silver label fatty gouaches, Maimeri experimented with an infinite number of substrates and managed to find the ideal one which, by its own physical nature, strongly imprisons the paint, allowing the perfectly ventilated layer of paint to react and to transform gradually from the state of soluble emulsion to the final state of insoluble film.

This result is due to a successful combination of a very fine top quality fabric with a mix suitable for ensuring proper absorbency for paint. On the right types of card, preparations of very fine, regular grain, it is possible to obtain <...>.

(c.29v.) CLOSE

I shall now say goodbye to my patient reader, with whom I have, in a spirit of simplicity, shared the fruit of my years of experience. I do not flatter myself by thinking that I have said things which are unique or entirely new, or provided (irrefutable) rules for obtaining good, solid paint.

Art is certainly something else, and no treatise or precept will ever serve to procure what can only be the fruit of personal inspiration. But making this easier, freeing up the field for artifices in this area so that all their work can be focused on expressing their own personality, cannot but be useful.

And so I must ask the reader's pardon if I have often had to make reference to a manufacturer in order to describe products, and the suspicion may be that I wish to advertise them. Certainly, several manufacturers, including a number of Italian ones, produce excellent materials with technical scruple and honesty. But I can guarantee the one I have mentioned, because I myself direct its technical management ever since it was established, and many of its products are precisely the fruit of the meditations I have <outlined> above for my colleagues in the hopes of being assistance to them.

Gianni Maimeri

Technical glossary

Luigi Brusati

Commodity name	**Lead White**	**Titanium White**
Names cited by Maimeri	*Silver white, Kremniz white, Biacca*	*Titanium white*
Former names and historical notes	Cerussa, cited and described by Theophrastus, Dioscorides, Vitruvius and Pliny.	Commercially available since 1913.
Chemical denomination	Lead(II) carbonate.	Titanium(IV) dioxide.
Mineralogical denomination	Hydrocerussite.	Exists in two crystalline forms, rutile and anatase, the former being of greater interest to the paints industry.
Chemical formula	$2Pb(CO)_3 . Pb(OH)_2$	TiO_2
Origin and preparation	One of the oldest artificial pigments, it was traditionally obtained through the action of vinegar fumes on lead sheets hung in suitable receptacles and kept in warm places. Its use greatly decreased after introduction of zinc white in the 19th century and titanium white in the 20th.	Obtained moist by hydrolyzing the clarified liquid stage got from a sulphuric attack on ilmenite ($FeTiO_3$).
Characteristics		
Colour and appearance	White.	White.
Colour index	PW 1	PW 6
Crystalline shape		Rhombohedral, trapezoidal.
Refraction indexes	Birefringent, $\varepsilon = 1.94$; $\omega = 2.09$	$\varepsilon = 3.256$; $\omega = 2.905$
Form of particles	Rounded.	
Specific weight, Kg/dm^3	6.14	Approx. 4.15
Mohs hardness		6.5
Solubility	In nitric acid.	In hot acid conc. and in melted alkalis.
pH	7.9	7.6
Oil absorption, g/100g pigment	8-15	18 - 22
Characteristic temperatures, °C	Decomposes at 400°.	
Notes	Toxic, blackens in contact with sulphides or sources of sulphur fumes.	Non-toxic, it is the most opaque pigment.

Zinc White	Lithopone
Zinc white	*Lithopone*
A white available since the late 18th century.	Created in the mid-19th century and also known as Griffith or Kneigt White.
Zinc(II) oxide.	Mixture of zinc sulphide (28-0%) and barium sulfate (70-72%).
Zincite.	
ZnO	$ZnS + BaSO_4$
Obtained by oxidizing zinc fumes, in the 19th century it replaced lead white in many uses.	Obtained moist, forming mixed precipitates by mixing solutions of zinc sulphide and barium sulphate.
White.	White.
PW 4	PW 5
Rhombohedral, trapezoidal.	
$\varepsilon = 3.256$; $\omega = 2.905$	Approx. 1.8
5.6	Approx. 4.3
In acids and alkaline solutions.	
7.6	7.6
10-22	14
Non-toxic, causes crackling in oil paints.	Light-sensitive.

Commodity name	**Azurite**	**Cobalt Blue**
Names cited by Maimeri	*Alemagna blues, copper carbonate*	*Cobalt blue, cobalt aluminate*
Former names and historical notes	Lazurium, lapis armenicum, azurim, azurium citramarinum, Magna blue.	Thenard Blue; the first preparations of cobalt enamels date from the first half of the 18th century, but the aluminate was commercially available in the early 19th.
Chemical denomination	Copper(II) carbonate.	Cobalt(II) aluminate.
Mineralogical denomination	Azurite.	
Chemical formula	$2Cu(CO)_3 . Cu(OH)_2$	indicatively $CoO . Al_2O_3$
Origin and preparation	One of the natural pigments used since antiquity, it could be obtained from the upper levels of copper deposits.	An artificial pigment obtained through the calcination of a mixture of cobalt (II) chloride and Al(III) chloride.
Characteristics		
Colour and appearance	Bright blue with greenish tones.	Blue.
Colour index	N/A	PB 28
Crystalline form	Monoclinic.	
Refraction indexes	Birefringent, $\alpha = 1.73$; $\beta = 1.758$; $\gamma = 1.838$	
Form of particles	Chips.	
Specific weight, Kg/dm³	3.77-3.83 (the purest)	
Mohs hardness	3.5-4	
Solubility	Soluble in acids and in ammonia.	
pH		
Oil absorption, g/100g pigment	23	45
Characteristic temperatures, °C	Decomposes at 220°.	
Notes	Colour intensity decreases with finer grinding.	

Cerulean Blue	Prussian Blue
Cobalt sky blue, cobalt stannate	*Prussian blue, basic ferrocyanide*
Höpfner introduced it in 1821 but it was hard to procure until 1860, when G. Rowney re-introduced it as a pigment more readily available. Currently still in use.	Berlin or Paris blue, first created in1714 by Diesbach in Berlin, made available in 1728 and still in use today.
Cobalt(II) stannate.	Iron(III) ferrocyanide(II).
indicatively $CoO\ .\ n\ SnO_2$	$Fe_3\ [Fe\ (CN)_6]_3$
An artificial pigment obtained by mixing cobalt(II) chloride and potassium stannate, afterwards treated chemically with silica and gypsum.	An artificial pigment. It forms as a precipitate from mixing potassium ferrocyanide solutions and soluble iron(III) salts.
Sky blue.	Deep blue.
PB 35	PB 27
	Cubic.
	1.83
	4.4
15-25	45
To be used in oil colours.	Toxic, unstable to light.

Commodity name	**Ultramarine Blue (artificial)**
Names cited by Maimeri	*Ultramarine, aluminum sulfo-silicate*
Former names and historical notes	From earliest times (approx. 6000 BC) it was made by grinding lapis lazuli. The resultant pigment was very costly but much admired for its colour and stability through time and it remained in use at least until the early 19th century. In 1787, during his travels in Italy, Goethe noted that in the Palermo area the blue crusts formed in limekilns were used as pigment. About 50 years later, in Saint Gobin, Tessaert noted the same crusts in soda kilns and had them analysed. Test results indicated the possibility of synthetically obtaining a pigment chemically similar to natural ultramarine blue. In 1824 the Societé d'Encouragement de l'Industrie Nationale offered a prize of 6000 francs to whoever managed to create a pigment similar to the natural one (at the time costing 4000 fr/lb.) at a much lower price. In 1828 J.B. Guimet achieved this and a blue pigment was created with the same components as the natural one.
Chemical denomination	Compound of sulphur and aluminum sodium silicate.
Mineralogical denomination	
Chemical formula	$Na_{6-8}Al_{5-6}Si_{6-7}O_{24}S_{2-4}$ The composition is not stoichiometric.
Origin and preparation	An artificial pigment that can be obtained through different procedures. One of them was, for example: mixing kaolin (100 p), sodium carbonate (70 to 82 p), sulphur (71 to 80 p) and charcoal (10 p) and firing it at 800°C. Afterwards selected on the basis of the colour of the masses obtained, ground and rinsed to remove the soluble sodium salts.
Characteristics	
Colour and appearance	Deep sky blue.
Colour index	PB 29
Crystalline shape	
Refraction indexes	Approx. 1.5
Form of particles	Rounded.
Specific weight, Kg/dm^3	Approx. 2.3
Mohs hardness	
Solubility	
pH	Approx. 9.4
Oil absorption, g/100g pigment	25-40
Characteristic temperatures, °C	
Notes	A stable colour but sensitive to acid substances.

Egyptian Blue	Smalt, Enamel Blue
Egyptian blue	*Mountain blue, cobalt silicate or enamel*
Used by the ancient Egyptians since about 3000 BC, afterwards prepared by the Romans in the Pozzuoli area and cited by Vitruvius and Pliny with the name "caeruleum". Also called Pompeii or Vestorian blue.	The first preparations date from the late 15th century but there are indications that cobalt-stained or similarly glazed glass was already known to the ancient Egyptians (27th century BC) in the Persian area.
Calcium and copper silicate.	Potassic glass with cobalt.
Similar to cuprorivaite.	
$CaCuSi_4O_{10}$	contains K, Si, As, Co
A glaze obtained by firing ($T > 800°C$) mixtures of silica, calcium carbonate, copper oxide or carbonate with sodium carbonate or borax to facilitate fusion.	A glaze composed of silica, potash and fired cobalt minerals (sulphides, sulphurarsenides and cobalt arsenates).
Light blue in variable shades.	Medium blue.
PB 31	PB 32
Tetragonal.	Amorphous.
$\alpha = 1.59$; $\omega = 1.633$	1.46-1.55
	Chips.
Approx. 3.3	
22	25

Commodity name	**Cadmium Yellow**
Names cited by Maimeri	*Cadmium yellow*
Former names and historical notes	Synthetic cadmium pigments became available in about 1848 after Stromeyer's discovery of cadmium in 1817.
Chemical denomination	Cadmium(II) sulphide mixed with variable amounts (4 to 15%) of zinc(II) sulphide crystals.
Mineralogical denomination	
Chemical formula	CdS (+ ZnS)
Origin and preparation	An artificial pigment made by treating solutions of cadmium salts with hydrogen sulphide or sulphurous alkaline solutions and thus obtaining a precipitate of cadmium(II) sulphide. After separation and desiccation it is heated to 500-600°C to transform its rhomboid crystalline shape to hexagonal. Pure cadmium sulphide is orange but can turn more yellow the more zinc sulphide there is in the mixed crystals.
Characteristics	
Colour and appearance	From pale yellow to golden yellow.
Colour index	PY 35
Crystalline shape	Hexagonal.
Refraction indexes	$\varepsilon = 2.506$; $\omega = 2.529$
Form of particles	
Specific weight, Kg/dm³	4.7-5
Mohs hardness	
Solubility	
pH	
Oil absorption, g/100g pigment	17-21
Characteristic temperatures, °C	Good heat resistance.
Notes	Does not last out of doors and is not suitable for murals.

Chrome Yellow(s)

Chrome yellow. Lead chromates ranging from chromatic acids to clearly basic chromates, from pale yellow to orange.

Chrome yellow is based on lead(II) chromate, prepared in the early 19th century by Vauquelin, after his 1797 discovery of chrome in a bright orange Siberian mineral consisting precisely of lead(II) chromate (crocoite).

Lead(II) chromate *(1)*. Also available commercially are pigments classified as chrome yellows *(2)*, having different tones of coloration and obtained by the presence of a range from $<5\%$ up to 50% of lead(II) sulphate. Chrome yellow with orange tones *(3)* also contains basic lead(II) chromate. Maimeri's description would seem to concern all these groups of pigments.

Crocoite.

(1) $PbCrO_4$; *(2)* $PbCrO_4 + PbSO_4$; *(3)* $PbCrO_4 . PbO$

An artificial pigment obtained by treating a solution of lead(II) nitrate with a solution of potassium chromate; the precipitate that forms is lead(II) chromate.

Yellow.

PY 34

Monoclinic.

$\alpha = 2.31$; $\beta = 2.37$; $\gamma = 2.66$

6.12

Approx. 7

25

Melts at 844°.

Toxic, it turns pale in the sun, blackens due to the sulphides and oxidises organic substances, for example organic colours.

Commodity name	**Zinc Yellow**
Names cited by Maimeri	*Zinc yellow*
Former names and historical notes	Obtained after Vauquelin's discovery of chrome in the late 19th century.
Chemical denomination	Zinc(II) chromate.
Mineralogical denomination	
Chemical formula	$4Zn.4CrO_3.K_2O.3H_2O$
Origin and preparation	An artificial pigment made by mixing a solution of a zinc salt with one of potassium chromate.
Characteristics	
Color and appearance	Lemon yellow.
Color index	PY 36
Crystalline shape	
Refraction indexes	
Form of particles	
Specific weight, Kg/dm³	
Mohs hardness	
Solubility	
pH	
Oil absorption, g/100g pigment	40
Characteristic temperatures, °C	
Notes	

Naples Yellow

Naples yellow, lead antimonate, antimonate yellow

A pigment known and prepared since antiquity by the Assyrians, Babylonians, Chaldeans and Persians and used to decorate pottery. It was found near the volcanic exhalations of Vesuvius.

Lead(II) antimonate.

$Pb(SbO_3)_2$ o $Pb(SbO_4)_2$

Also found as a mineral, but an artificial pigment was made through the calcination of lead compounds (oxide, carbonate, etc.) mixed with antimonate compounds (oxide, sulphide, etc.).
It can also be made moist, mixing a solution of lead(II) nitrate with one of potassium antimonate.

Orangey yellow.

PY 41

2.01/2.28

4.6/4.7

10/15

Highly toxic, it blackens in contact with sulphur fumes and sulphides.

Commodity name	**Yellow Ochre (natural)**	**Red Ochre (natural)**
Names cited by Maimeri	*"Ochres and colored earths"*	*"Ochres and colored earths"*
Former names and historical notes	A pigment used since prehistoric times, for example in the Lascaux cave paintings.	A pigment used since prehistoric times.
Chemical denomination	Yellow ochre is a mixture whose main components are kaolin clays, silica and hydrated iron(III) oxide, which accounts for the coloration.	Red ochre is a mixture whose main components are kaolin clays, silica and anhydrous iron(III) oxide, which accounts for the coloration.
Mineralogical denomination	α-Göethite (hydrated iron(III) oxide).	Hematite.
Chemical formula	FeO(OH) or $Fe_2O_3.H_2O$	α-Fe_2O_3
Origin and preparation	A natural pigment that varies in color.	A natural pigment that varies in color.
Characteristics		
Color and appearance	Hydrated iron(III) oxide in the microcrystalline form is yellow.	Red.
Color index	N/A	PR 101:1
Crystalline shape	Rhombohedral.	Hexagonal.
Refraction indexes		
Form of particles		
Specific weight, Kg/dm³	2.8-3.5	
Mohs hardness		
Solubility		
pH		
Oil absorption, g/100g pigment	35	15-25
Characteristic temperatures, °C	Between 350° and 400° it loses water, progressively transforming into red iron(III) oxide.	
Notes		

Green Earth	Cassel Earth
"Ochres and colored earths"	*"Ochres and colored earths"*
A pigment used since prehistoric times.	Also called Cologne Earth, Van Dyke Brown and in use since the late 17th century.
Aluminum silicate compounds.	Lignite or moist peat, hydrated iron(II) oxide, partially hydrated manganese(IV) dioxide.
Celadonite and glauconite.	Contains lignite, göethite, pyrolusite.
$K[Al\ Fe^{III}, Fe^{II}, Mg](Al\ Si_3, Si_4)\ O_{10}\ (OH)_2$	$Fe_2O_3.H_2O, MnO_2.nH_2O$
A natural pigment.	Originates from a mineral deposit of the earthy-peat type. After desiccation, depuration and grinding it yields a natural pigment.
Green.	Dark brown.
PG 23	PBr 8
Prismatic monoclinic.	
$\alpha = 1.39\text{-}1.612$; $\beta = 1.609\text{-}1.643$: $\gamma = 1.61\text{-}1.644$	
20-30	

Commodity name	Umber (natural)	Umber (burnt)
Names cited by Maimeri	*"Ochres and coloured earths"*	*"Ochres and coloured earths"*
Former names and historical notes	A pigment known since earliest times, whose name apparently derives from Umbria, where it was quarried.	A pigment known since earliest times, whose name apparently derives from Umbria, where it was quarried.
Chemical denomination	This earth is a mixture of minerals whose main component is partially hydrated iron(III) oxide, with fair amounts (up to 20%) of manganese (IV) dioxide and also organic substances.	This earth is a mixture whose main component is anhydrous iron(III) oxide with fair amounts of manganese(IV) dioxide.
Mineralogical denomination		
Chemical formula	$nFe_2O_3 . H_2O; MnO_2$	$Fe_2O_3; MnO_2$
Origin and preparation	A natural pigment.	A natural pigment subjected to red heat. This eliminates hydration water and organic substances, making the colour deeper than the natural earth.
Characteristics		
Colour and appearance	Dark brown.	Deep dark brown.
Colour index	PR 7/PBr 8	PBr 7
Crystalline shape		
Refraction indexes		
Form of particles		
Specific weight, Kg/dm³	3.48	3.48
Mohs hardness		
Solubility		
pH		
Oil absorption, g/100g pigment	44	60-80
Characteristic temperatures, °C		
Notes		

Siena (natural)	Siena (burnt)
"Ochres and coloured earths"	*"Ochres and coloured earths"*
Siena is a pigment used since earliest times and named for the place where it was quarried.	Siena is a pigment used since earliest times and named for the place where it was quarried.
This earth is a mixture whose main component is partially hydrated iron(III) oxide with traces of manganese(IV) dioxide.	This earth is a mixture of minerals whose main component, after calcination, is iron(III) oxide with traces of manganese(IV) dioxide.
$nFe_2O_3.H_2O$	Fe_2O_3
A natural pigment.	
Yellow-brown.	Red-brown.
PR 6	PR 6
3.46	3.80
38	28-43

Commodity name	**Cinnabar**
Names cited by Maimeri	*Cinnabar, vermilion*
Former names and historical notes	A pigment known from the earliest times and also man-made in the Orient from very early times. Pliny cited as "minium" the kind quarried in Spain. Cinnabar stood for the compound of natural origin, vermilion for the artificial.
Chemical denomination	Mercury(II) sulfide (α) Cadmium(II) sulfide + Cadmium(II) selenide
Mineralogical denomination	Cinnabar.
Chemical formula	HgS α-$CdS + CdSe$
Origin and preparation	Mineral-origin cinnabar was simply ground and rinsed, while the artificial was made by sublimation of a mixture of mercury and sulphur. When mixed, mercury and sulphur create a black compound. Long ago a great amount of sulphur was used, while between the 16th and 17th centuries the ratio was nearly stoichiometric. In the late 18th century Kirchoff perfected a process of preparing cinnabar wet.
Characteristics	
Colour and appearance	Orangey red.
Colour index	PR 106
Crystalline shape	Rhombohedral.
Refraction indexes	$\omega = 2.905$; $\varepsilon = 3.256$
Form of particles	
Specific weight, Kg/dm³	8.1
Mohs hardness	
Solubility	
pH	
Oil absorption, g/100g pigment	7-12
Characteristic temperatures, °C	Sublimates at 583.5°.
Notes	

Cadmium Red

Cadmium red or cadmium sulphurselenide

Available since the 1830s.

An artificial pigment in which the yellow of cadmium sulphide is changed from orange to bright red due to the presence of selenide in variable amounts from 25 to 90%. It can be prepared dry and moist, the latter by precipitating with sodium sulphide from a cadmium sulphate solution in which selenide is present.

From orange to bright red.

PR 108

7-8.5

17-21

Commodity name	Robbia and Alizarin Lake
Names cited by Maimeri	*Madder lake, alizarin lake*
Former names and historical notes	Before the arrival of a similar, man-made colourant (1868), this was obtained from the roots of the *Rubia tinctorum*, a plant in the Rubiacee family, whose cultivation was widespread from ancient times. In the roots of the young plant nearly twenty molecular species with colouring power have been identified, while in the mature plants they are greatly reduced and mostly contain alizarin and purpurin (1,2,4 - trihydroxyanthraquinone and a few other compounds, always with an anthraquinonic structure.
Chemical denomination	(natural), alizarin + purpurin; (synthetic) only alizarin.
Mineralogical denomination	
Chemical formula	(natural) 1,2 - dihydroxyanthraquinone + 1,2,4 - trihydroxyanthraquinone; (synthetic) 1,2 - dihydroxyanthraquinone.
Origin and preparation	Like almost all colourants of vegetal origin, anthraquinone compounds are found in nature in the form of glucosides, or tied to sugar molecules that increase their solubility in watery mediums. The latter have to be eliminated to permit the full development of colouring power, using processes of fermentation. The anthraquinonic compounds of th Robbia remain water-soluble even after being freed of natural glucoside derivatives. The are sensitive to the pH in the watery medium used for preparation and use and to the type of metallic ions present, and they can also be prepared in the form of lakes, making them absorb on mineral supports. For example, by mixing an alum solution (aluminum and potassium sulphate) with the colouring solution and precipitating the aluminum hydroxide by adding alkaline solutions. The colourant sets on the aluminium, thus forming a red lake. Also usable are other absorbent minerals as Maimeri suggests, but aluminium is the best.
Characteristics	
Colour and appearance	Variable red.
Composition	
Colour index	NR 8 - NR 12
Crystalline shape	
Refraction indexes	
Form of particles	
Specific weight, Kg/dm³	Approx. 1.7
Mohs hardness	
Solubility	
pH	
Oil absorption, g/100g pigment	80 (NR 9)
Characteristic temperatures, °C	
Notes	

Synthetic Iron Oxides

In a class by themselves are the so-called synthetic earth oxides, which range from pale yellow to brown. Ochres and coloured earths are purplish to red. They are very solid colours with great covering power.

These artificial pigments consist of iron (II) and (III) oxides, some hydrated or anhydrous. They have iron content expressed as a % of Fe_2O_3 varying from about 88 to 94, with impurities.

α-Göethite, hematite, magnetite.

See table.

Yellow.	Orange.	Brown.	Red.	Black.
α-FeO(OH)	γ-FeO(OH)	γ-Fe_2O_3	α-Fe_2O_3	FeO. Fe_2O_3
PY 42		PBr 6	PR 101	N/A
Rhombohedral.			Hexagonal.	Magnetite.
3.7-4.3	3.7-4.1	4.2-4.8	4.4-5.3	4.5
3.5-4	6-6.5	3.5-4	5-5.8	Approx. 4.5
20-30	Approx. 30	28-43	15-25	15-25
Between 350° and 400° it loses water, gradually turning into red iron(III)-oxide.	Between 350° and 400° it loses water, gradually turning into red iron(III)-oxide.	Melts at approx. 1595°.	Melts at approx. 1595°.	
			Impure.	Impure.

Commodity name	Chrome Oxide Green	Hydrated Chrome Oxide Green
Names cited by Maimeri	*Chrome oxide or grass green*	*Guignet or emerald green, chrome oxidrate*
Former names and historical notes	Prepared by Vauquelin in 1809, first used for pottery glazes and later as a painting pigment.	Viridian; prepared in 1838 and patented in 1859 by Guignet.
Chemical denomination	Chrome(III) oxide.	Dihydrate chrome(III) oxide.
Mineralogical denomination		
Chemical formula	Cr_2O_3	$Cr_2O_3.2H_2O$
Origin and preparation	An artificial pigment made through the calcination of alkaline dichromates in the presence of a reducing agent, for example potassium dichromate with carbon.	An artificial pigment made through the calcination, at 500°C, of an alkaline dichromate with boric acid. The resultant compound, treated in water, is a green precipitate, the pigment desired.
Characteristics		
Colour and appearance	Olive green.	Green.
Colour index	PG 17	PG 18
Crystalline shape	Monoclinic.	
Refraction indexes	$\alpha = 1.62$; $\beta = 2.12$	$\alpha = 1.71/2$; $\beta = 1.77/8$
Form of particles		
Specific weight, Kg/dm³	5.09	3.71
Mohs hardness		
Solubility		
pH		
Oil absorption, g/100g pigment	12-43	45-90
Characteristic temperatures, °C	Resistant up to 1000°.	At 250° begins to transform into chrome(III) oxide.
Notes		

Cobalt Green	Schweinfurt Green
Cobalt greens, which are combinations of zinc and cobalt	*Schweinfurt greens or Paolo Veronese green*
	A toxic pigment now in disuse, perfected in 1808 by Scheele and made in 1814 by Sattler in Schweinfurt.
Zinc(II) chromate + Cobalt(II) aluminate.	Copper(II) acetoarsenite.
$4Zn.4CrO_3.K_2O.3H_2O + CoO . Al_2O_3$	$Cu(CH_3COO)_2.3Cu(AsO_2)_2$
An artificial pigment made by mixing a yellow pigment with a blue one.	An artificial pigment obtained through heated mixture of copper (II) sulphate with an arsenite potassium. Acetic acid is added to the voluminous precipitate that forms.
Green from yellowish to bluish.	Emerald green.
PG 26	PG 21
	Approx. 4
20	10

Commodity name	Ivory Black (Bone Black)
Names cited by Maimeri	*Ivory black*
Former names and historical notes	Ivory black, bone black; pigments of extremely early origin.
Chemical denomination	It is a mixture of calcium phosphate, calcium carbonate and carbon, indicatively in percentages of 84, 6 and 10.
Mineralogical denomination	
Chemical formula	$Ca_3(PO_4)_2 + Ca(CO)_2 + C$
Origin and preparation	An artificial pigment obtained by carbonising discards from working ivory, and quite often degreased animal bones.

Characteristics	
Colour and appearance	Black.
Colour index	PBk 9
Crystalline shape	
Refraction indexes	
Form of particles	
Specific weight, Kg/dm³	Approx. 2.5
Mohs hardness	
Solubility	
pH	
Oil absorption, g/100g pigment	35-45
Characteristic temperatures, °C	
Notes	

Sepia Black	Vine Black
Sepia: a natural product made from cuttlefish ink, with good solidity	*Vine black, vegetal black*
	Like ivory black and bone black it is a pigment of extremely early origin.
Melanin, which is an amorphous biopolymer of indole quinone.	Carbonic substances.
A natural pigment taken from cuttlefish ink, whose characteristic constituent is melanin.	An artificial pigment made by carbonising pruned vine shoots (see Cennini, *Il libro dell'arte/The Book of Art*). Later versions were made by carbonising wine dregs and grape stalks.
Black with brown tones.	Black.

Commodity name	**Cobalt Violet**
Names cited by Maimeri	*Cobalt violet, cobalt phosphate, prepared in two shades, light and dark*
Former names and historical notes	The anhydrous cobalt(II) phosphate prepared by Salvètat in 1859 is the deeper cobalt violet while the cobalt(II) arsenate prepared in Germany is lighter in colour.
Chemical denomination	Cobalt(II) phosphate or cobalt(II) arsenate.
Mineralogical denomination	
Chemical formula	$Co_3(PO_4)_2$ or $Co_3(AsO_4)_2$
Origin and preparation	An artificial pigment obtained as a precipitate by mixing a solution of cobalt (II) chloride with one of sodium bisulphate (NaH_2PO_4).

Characteristics	They refer to cobalt(II) phosphate.
Colour and appearance	Violet.
Colour index	PV 14
Crystalline shape	
Refraction indexes	
Form of particles	
Specific weight, Kg/dm³	2.59
Mohs hardness	
Solubility	
pH	
Oil absorption, g/100g pigment	16-24
Characteristic temperatures, °C	
Notes	

Manganese Violet

Mineral violet, manganese phosphate

Permanent violet introduced in the mid-19th century.

Ammonium manganese (III) biphosphate + manganese(II) metaphosphate.

$(NH_4)_2Mn_2(P_2O_7)_2 + Mn_3(PO_4)_2$

An artificial pigment, prepared more elaborately than many other pigments. First, solutions of manganese(II) chloride and phosphoric acid are mixed, evaporated and dried. The dry part is heated until it melts. This mass is broken down and rinsed by boiling with an ammonium-carbonate solution. The sediment is re-melted and then rinsed in boiling water. The insoluble part consists mainly of manganese(II) metaphosphate.

Violet.

PV 16

Approx. 25

Commodity name	**Turpentine**
Names cited by Maimeri	*Turpentine*
Former names and historical notes	
Chemical characteristics	Its main components are a mixture of monoterpenoids such as pinene and limonene.
Origin and preparation	Obtained from the steam-current distillation of different types of pine, less frequently of larch and fir.

Characteristics	
Appearance	Liquid, with characteristic odour.
Specific weight, Kg/dm³	0.86-0.87
Refraction indexes	1.463-1.470
Boiling range, °C	152-160
Solubility	Complete in pure ethanol, aromatic hydrocarbons, chlorides, etc.
Notes	Leaves some resinous residue.

Lavender essence oil	White Spirit
Lavender essence	*Petroleum essence, petroleum ether, gasoline, white spirit.*
Began to be used in the 16th century.	
A complex mixture of monoterpenoids, both hydrocarbons and their oxygenated derivatives, also with some sesquiterpenoids. Linalool and linalyl acetate have been identified as major components.	A mixture of paraffin and cycloparaffin hydrocarbons and some aromatic alkalines.
Obtained from the steam-current distillation of *Lavandula spica*, which differs from the *Lavandula fragrans* used in perfumery. Composition varies, depending on the geographical zones of origin.	A petroleum distillate.
Dense liquid with characteristic odour.	Liquid.
0.89-0.91	0.77-0.79
1.457-1.469	1.41-1.44
150-250	150-220
In ethanol.	
When evaporated leaves some resinous residue.	Leaves no residue.

Commodity names:	**Linseed, walnut and poppy seed oils**

Chemical characteristics: Natural fats such as these vegetal oils consist of glycerin tri-esters (triglycerides) with characteristic mixtures of medium-long chain aliphatic fatty acids. These aliphatic chains may or may not have double bonds (called unsaturated) between the carbon atoms comprising them. In fatty acids more significant than these oils the double bonds can vary from 1 to 3 per aliphatic chain; respectively, 1 for oleic acid, 2 for linoleic acid and 3 for linolenic acid. The latter is more reactive to oxygen in the air and its content mainly depends on the drying characteristics of the oil.

Origin and preparation: These oils are traditionally extracted by pressing the plant parts that contain them, but they can also be processed with solvents. The crude oils undergo various treatments to improve limpidity, color, smell and consistency and to reduce acidity, generally improving purity.
Linseed oil is also given treatments to increase consistency and boost drying power, heating it in the presence of oxygen or using lead or other metal compounds (heat-treated linseed oil, with drying agents, etc.).

The following table summarizes the most common characteristics of the oils Maimeri discusses in his manuscript.

Characteristics	**Linseed Oil**
Appearance and colour	Limpid; from yellow to reddish and brownish yellow.
Physical characteristics	
Specific weight at 15°, Kg/dm³	0.930-0.935
Freezing range, °C	-18° / -27°
Refraction index range	1.479-1.481
Compositional characteristics	
Number of iodine[1] mg of iodine/g of oil	169-172
Amount of saponification, mg KOH/g of oil[2]	190-195
Saturated fatty acids, %[3]	9-13
Unsaturated fatty acids, %	
– oleic acid	14-24
– linoleic acid	14-19
– linolenic acid	48-60

NOTES:
1. Iodine is added to the unsaturated bonds of the aliphatic chain (2 atoms every double bond) and so the amount in mg added per g of oil indicates the average amount of these bonds present, whatever the type of acid.
2. The amount of saponification is an index in reverse proportion to the average molecular weight of the fatty acids, assuming that each acid has just one acid function per aliphatic chain.
3. The compositions given here are indicative due to the variability of natural products; they are the result of gas-chromatography tests that may differ perceptibly from past data traditionally used.

Walnut Oil	Poppy Seed Oil
Limpid; pale yellow.	Limpid, golden yellow.
0.925-0.929	0.923-0.926
-15° / -18°	-15° / -20°
1.469-1.475	1.475-1.478
140-161	131-143
189-197	189-197
4-10	12
9-30	11
57-76	72
2-16	5

Commodity names:	**Egg** ***(albumen or egg white; yolk)***

Compositional characteristics: The main constituents, with indication of their percentages, are:

	albumen (approx. 58% of a whole egg)	**yolk** (approx. 32% of a whole egg)
Water %	Approx. 85	Approx. 50
Albumin and other proteins, %	Approx. 13	
Proteins, %		Approx. 16
Lipid substances (triglycerides) %	< 0,3	Approx. 25
Phosphatides (including lecithin), %		7-10
Cholesterol, %		Approx. 1.5
Minerals, %	< 1	Approx. 1
Indicative composition of the fatty acids in the yolk's lipids		(gas-chromatographic methods)
– saturated acids %		36
– oleic acid, %		44
– linoleic acid, %		13-14
– linolenic acid, %		0.5-1

NOTES:

Albumen: this was used especially in illuminated manuscripts. The utility of albumen in various preparations was primarily tied to its binding power. However, to make it suitable for use it had to be rendered liquid, mechanically destroying its gelatinous structure. This was done by beating it with sheaves of twigs or with sponges. After the foam disappeared the liquefied albumen could disperse pigments and dissolve colorants.

Yolk: this contains less water than the albumen (see table), which holds in emulsion the part that can be called "dry" (lipids, phosphatides, proteins, cholesterol).

Worthy of note is the stability of the emulsion thanks to its cholesterol and lecithin, making it easy to disperse pigment until it reaches the right consistency for painting. Lipid composition features the presence of a fair amount of unsaturated fatty acid esters with drying power.

Before oil paints became common many old masters used egg yolk as a binder (tempera painting).

Plates

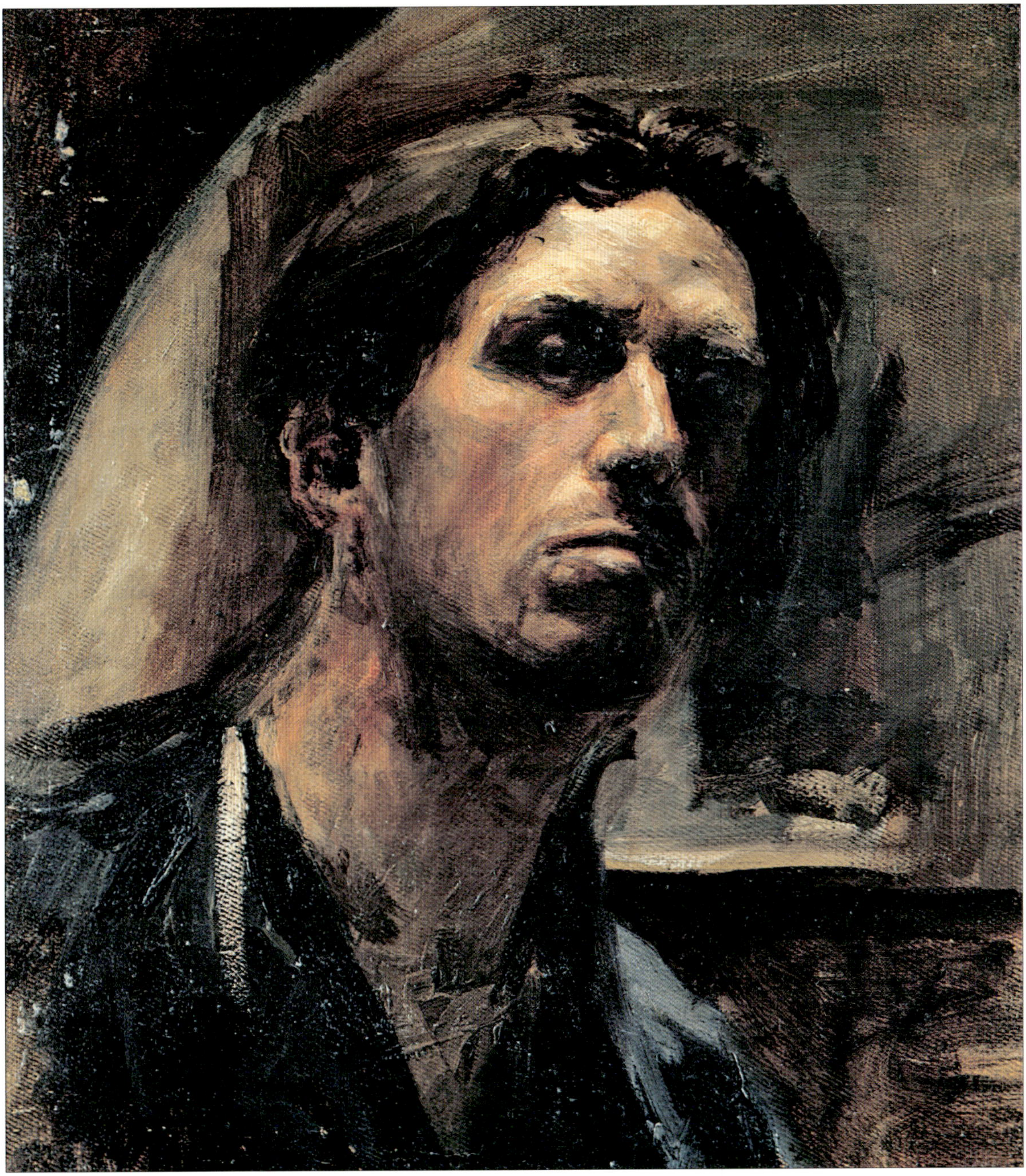

1. "AUTORITRATTO GIOVANILE", 1916. OIL ON CANVAS, 50 X 40.5 CM, DATED AND UNSIGNED.

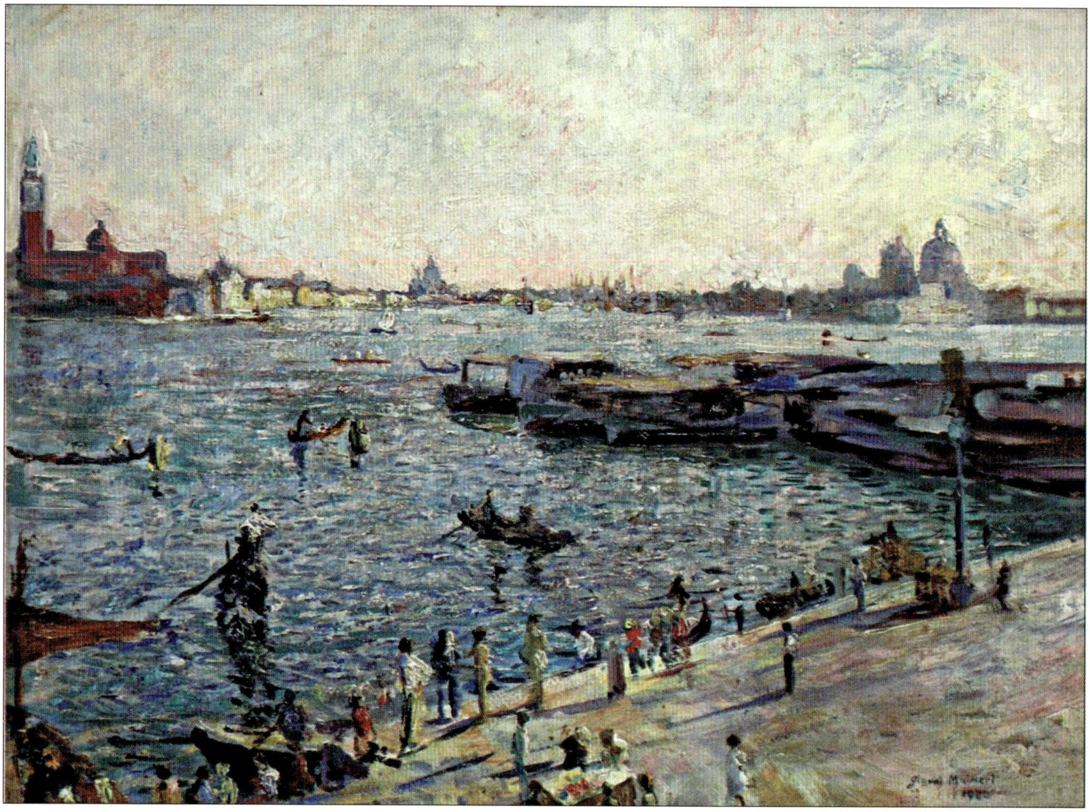

2. "Il bacino di San Marco", 1920. Oil on cardboard, 50 x 65 cm, dated and signed.

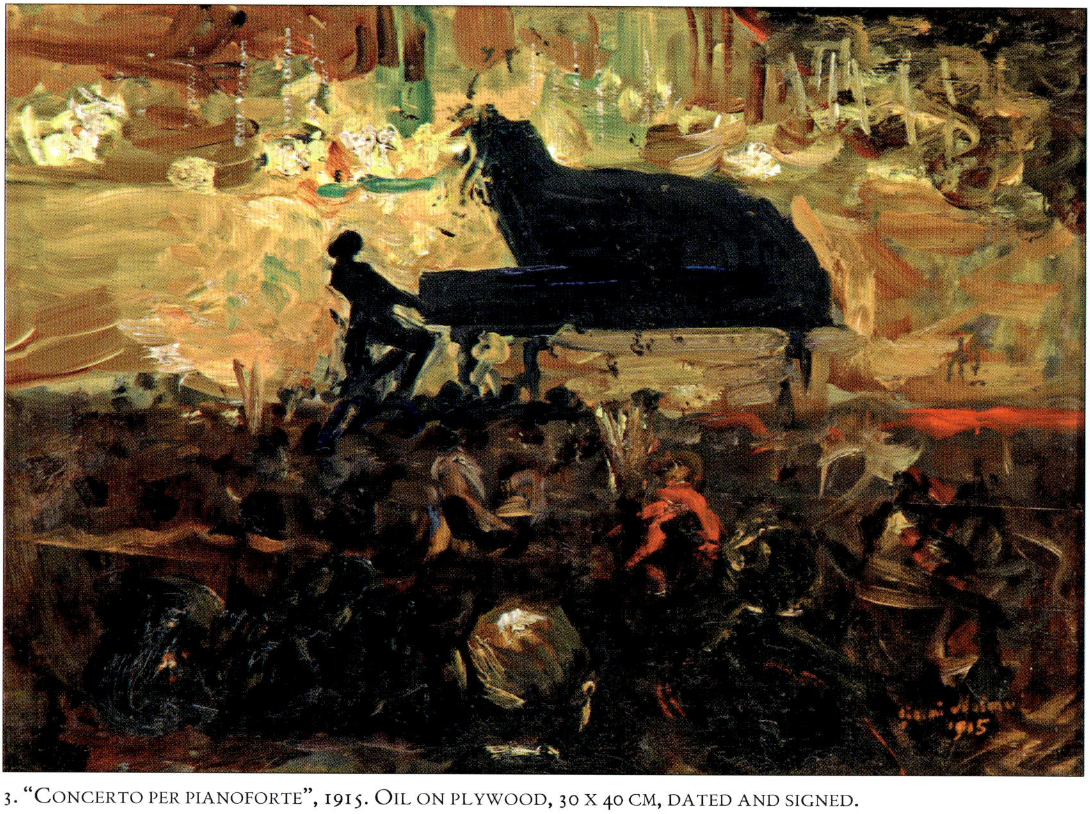

3. "Concerto per pianoforte", 1915. Oil on plywood, 30 x 40 cm, dated and signed.

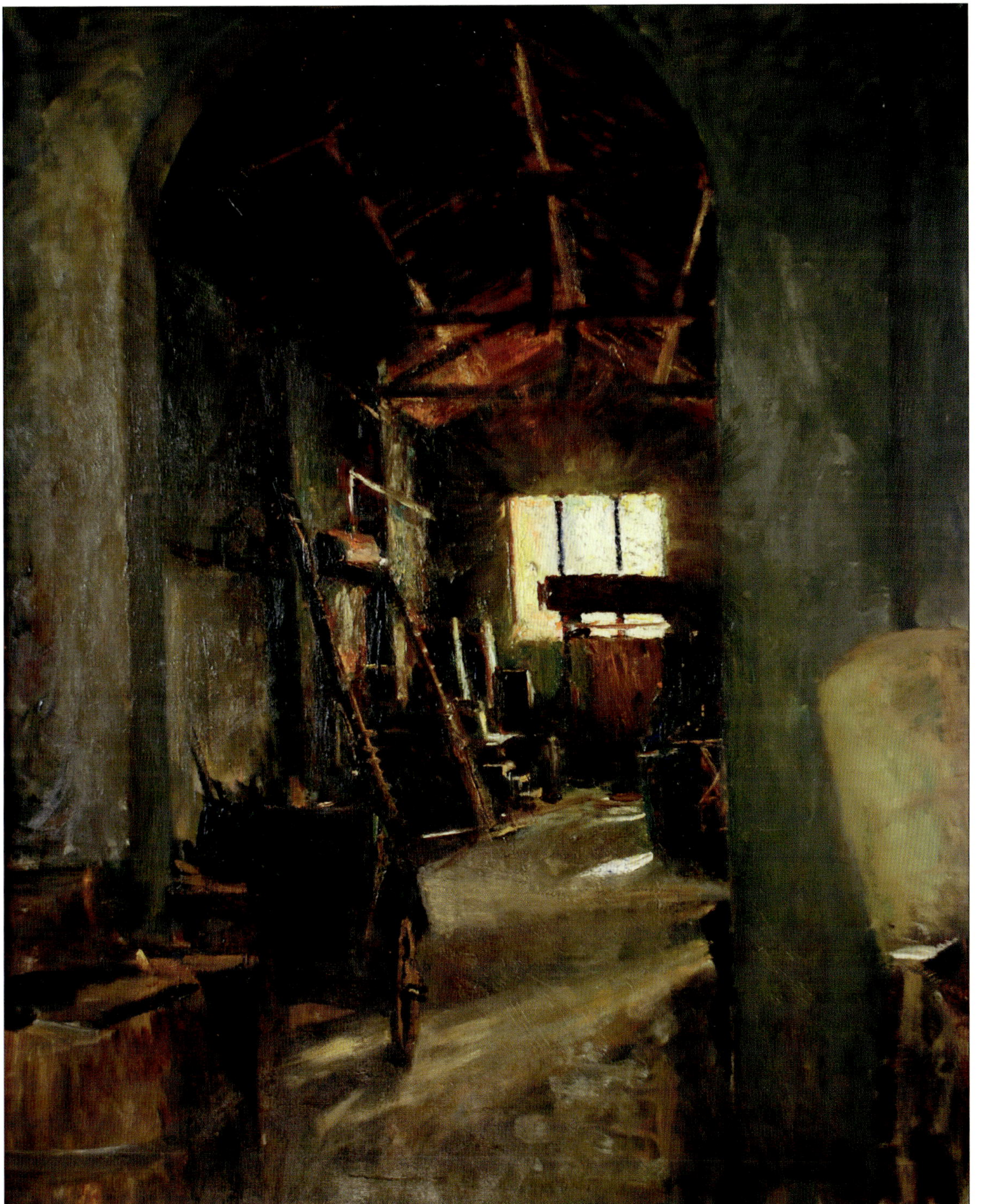

4. "Interno dello stabilimento Maimeri alla Barona", 1929 (?).
Mixed technique on canvas, 102 x 83 cm, unsigned.

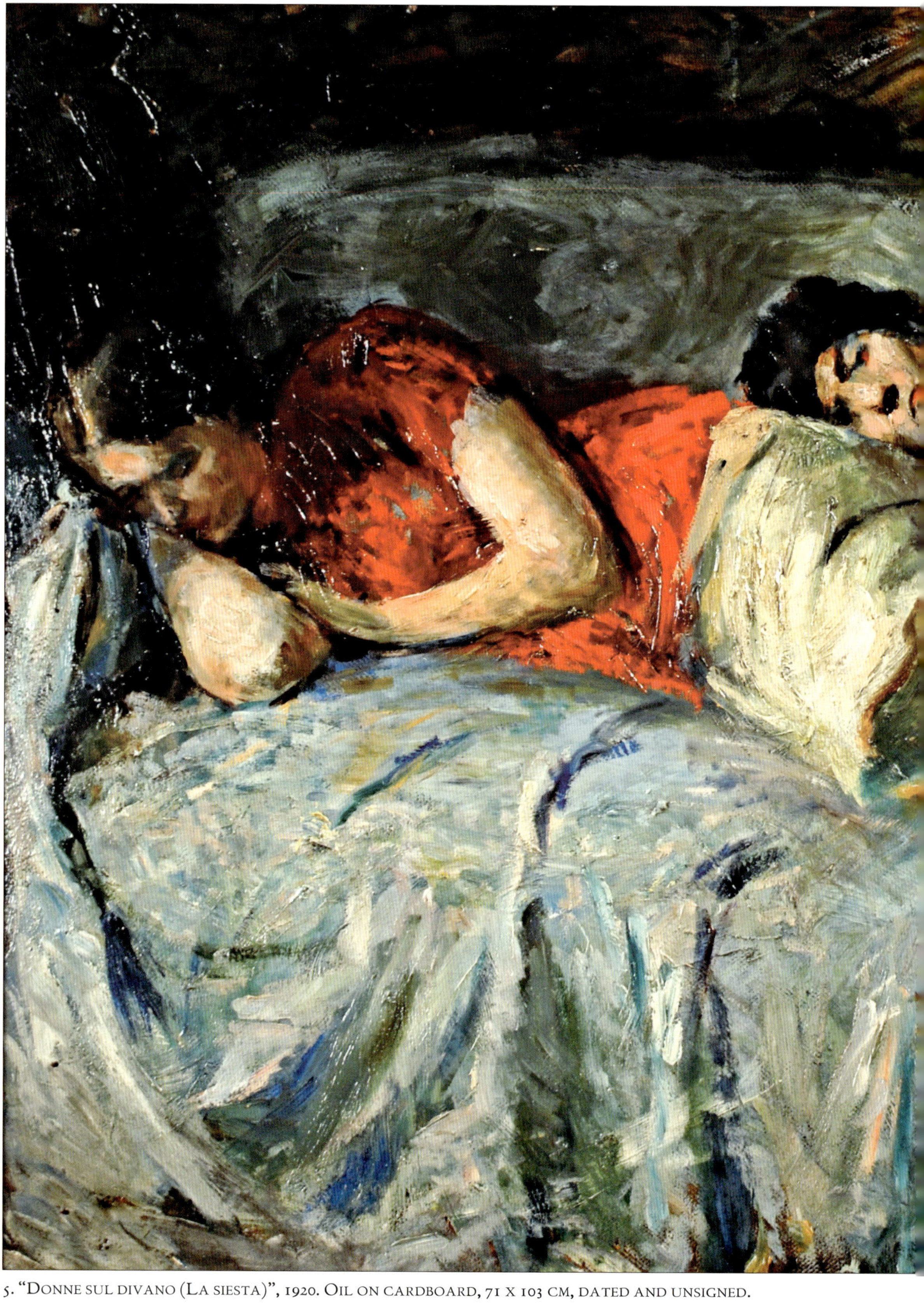

5. "Donne sul divano (La siesta)", 1920. Oil on cardboard, 71 x 103 cm, dated and unsigned.

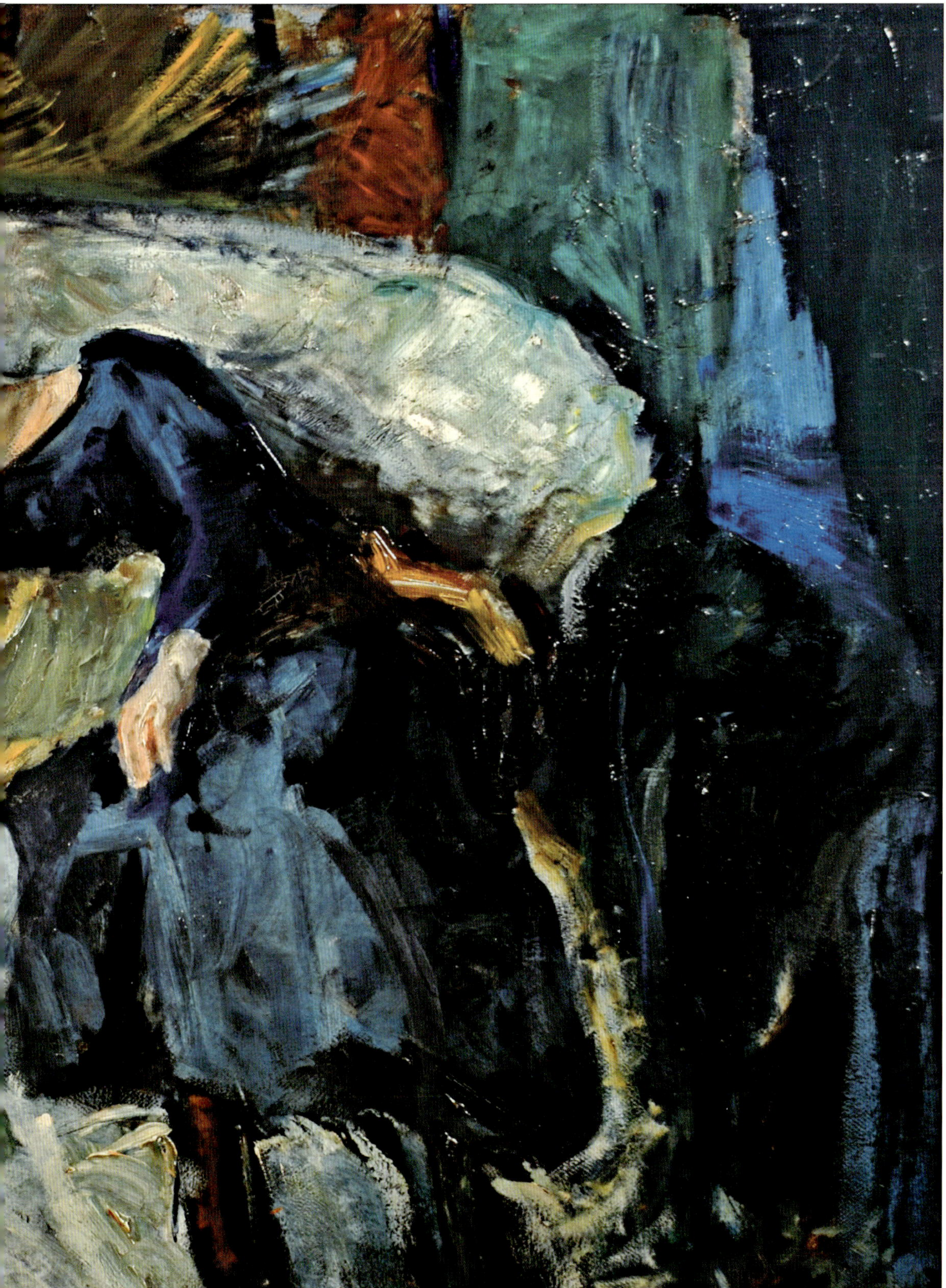

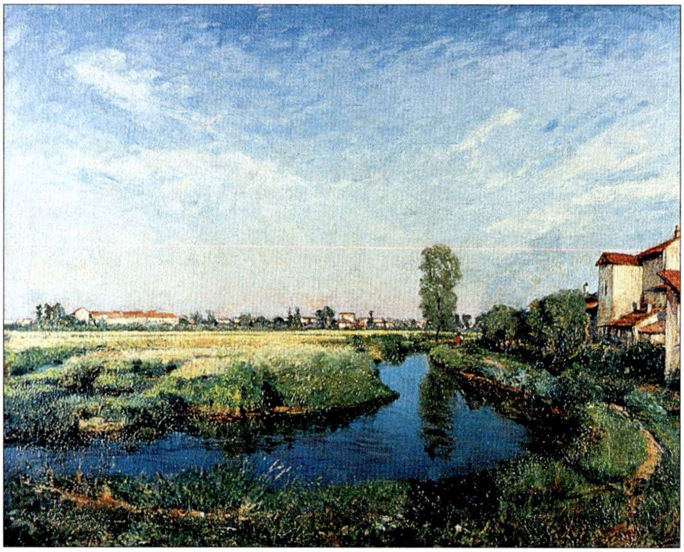

6. "L'ESTATE", 1920.
OIL ON CANVAS, 120.5 X 151 CM,
DATED AND SIGNED.

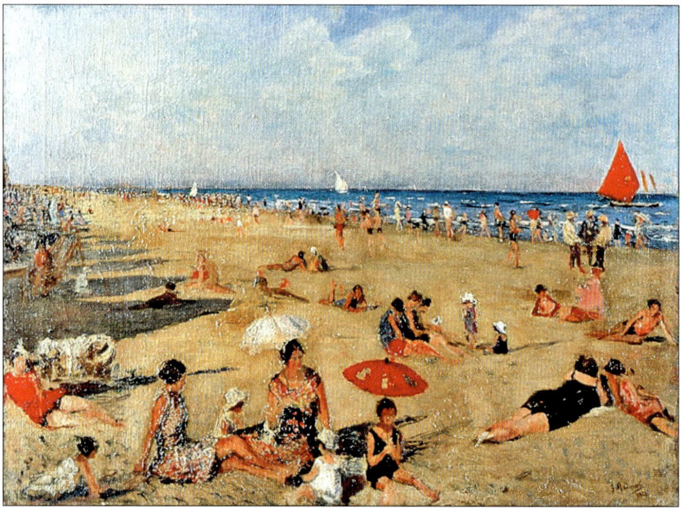

7. "LIDO DI VENEZIA", 1923.
OIL ON CANVAS, 90 X 120 CM,
DATED AND SIGNED.

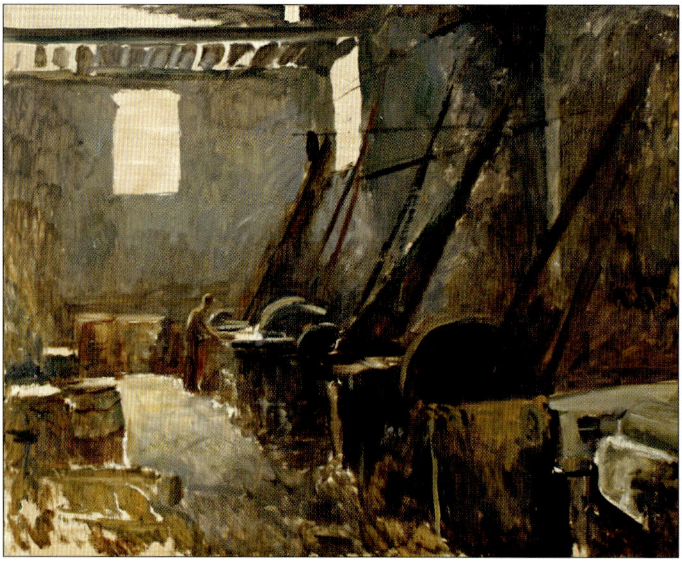

8. "INTERNO DELLO STABILIMENTO
MAIMERI ALLA BARONA", 1929.
OIL ON CANVAS, 150 X 120 CM, UNSIGNED.

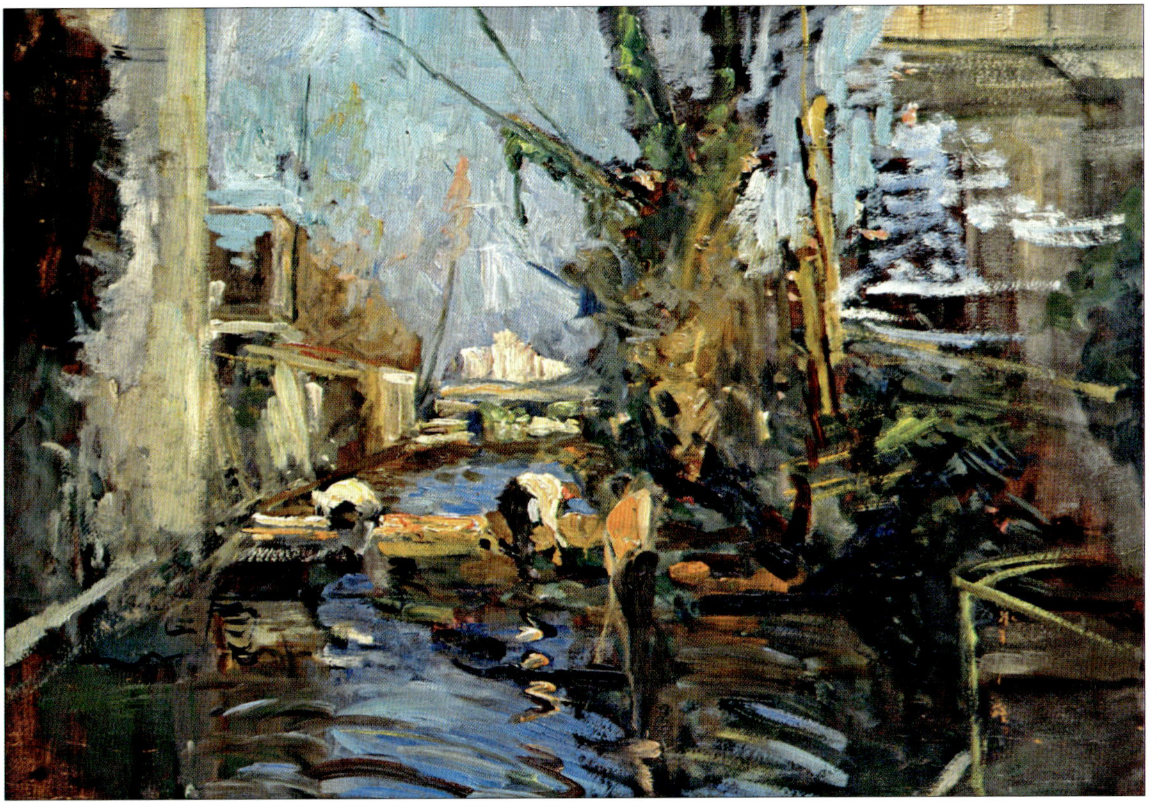

9. "Le lavandaie", 1915. Oil on cardboard, 52 x 72 cm, unsigned.

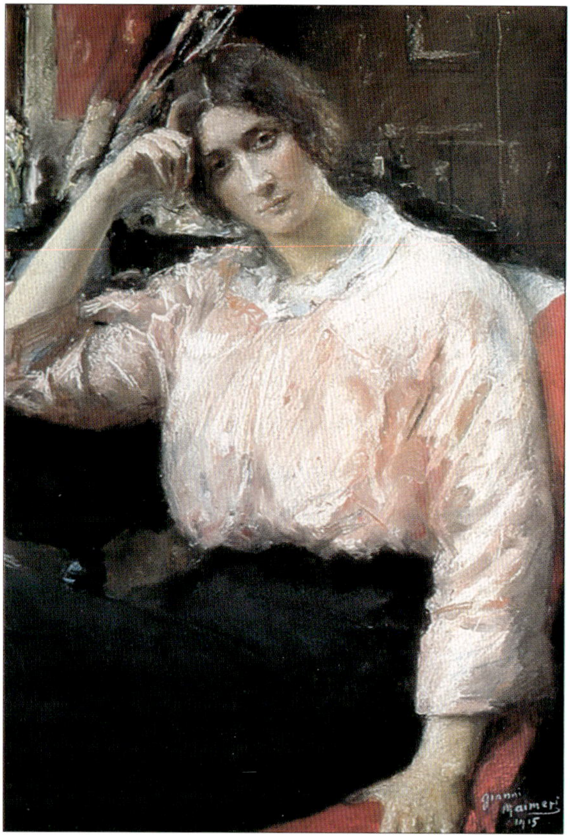

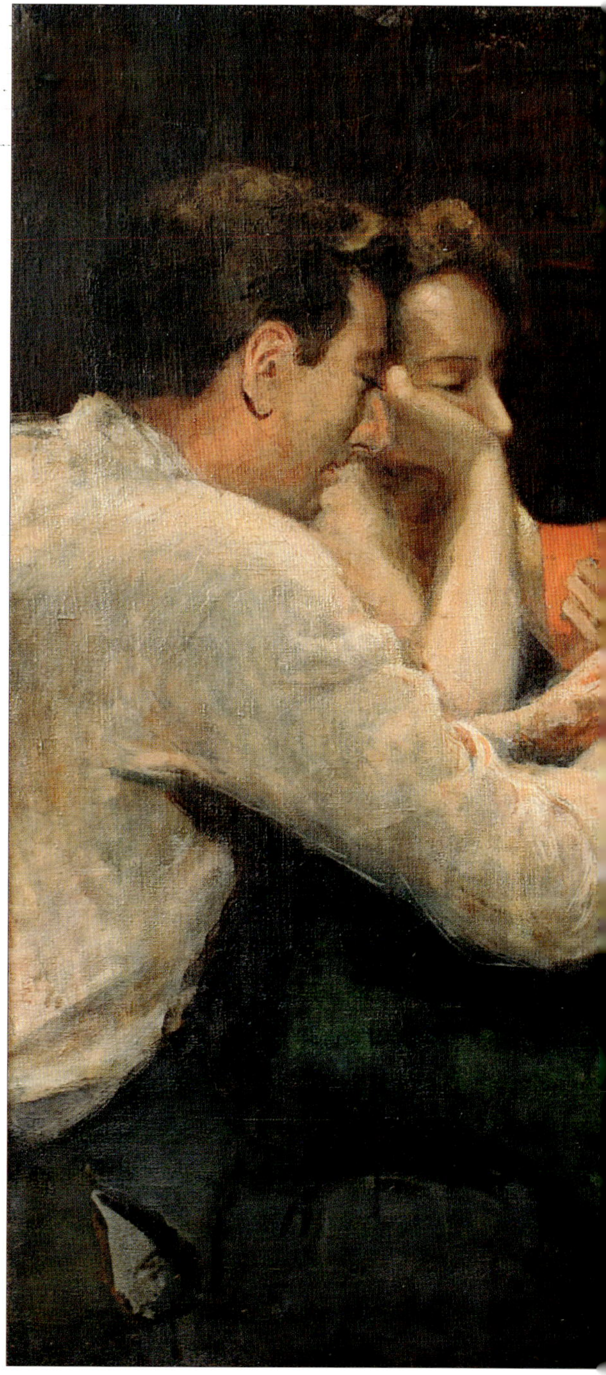

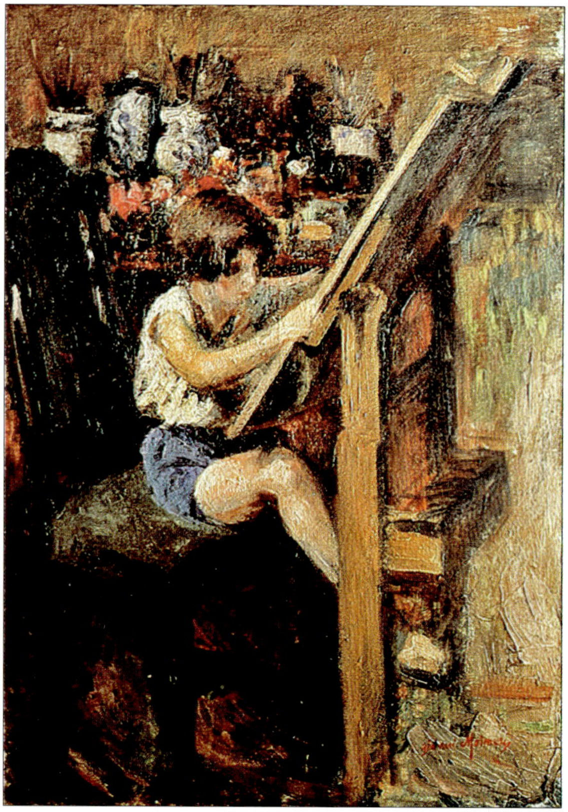

10. "Mia moglie", 1915.
Pastel on cardboard, 90 x 65 cm,
dated and signed.

11. "Mio figlio al cavalletto",
1933 (?). Oil on canvas, 110 x 67 cm,
dated and signed.

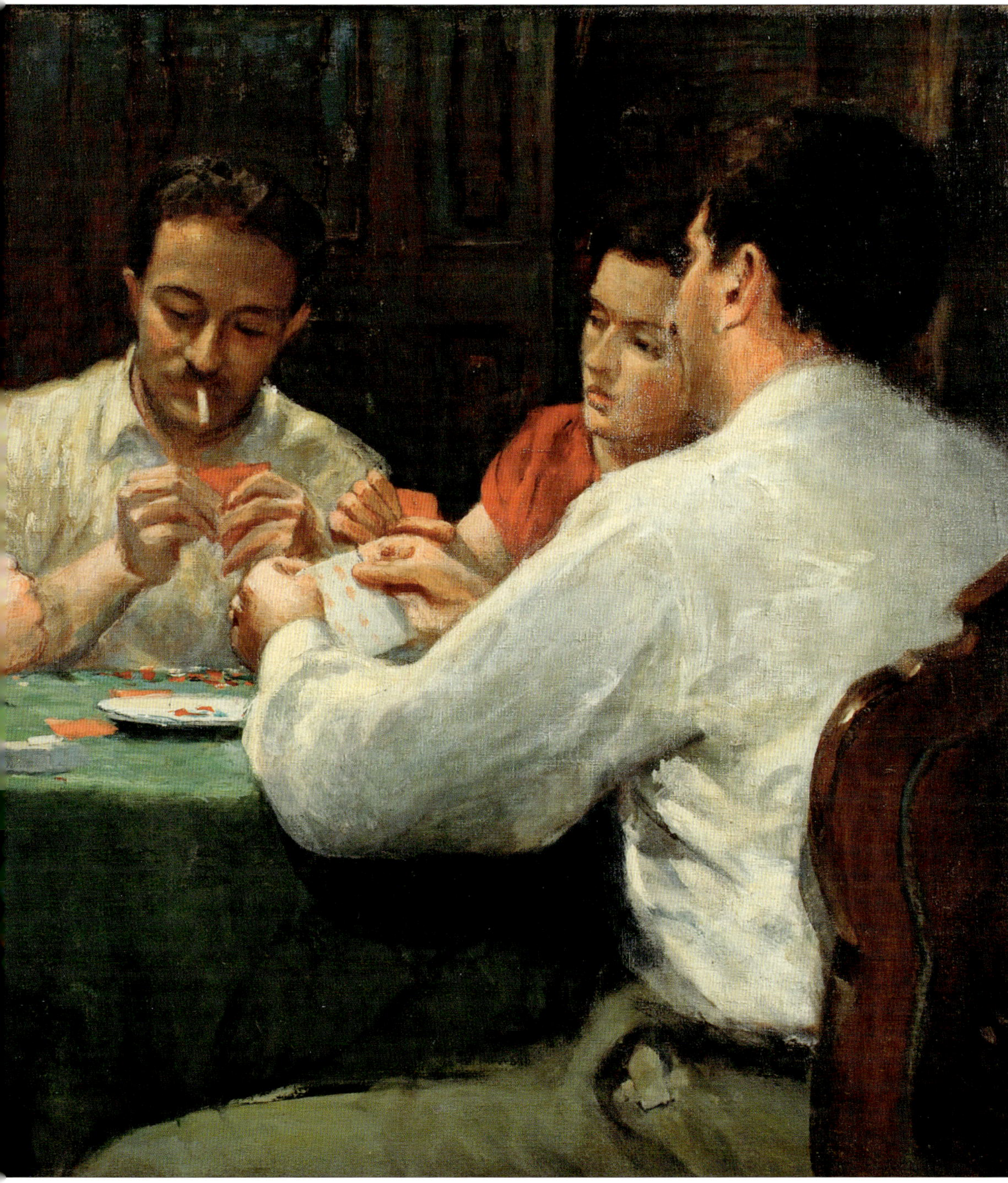

12. "Partita a poker", 1951 (unfinished). Tempera on canvas, 90 x 120 cm, unsigned.

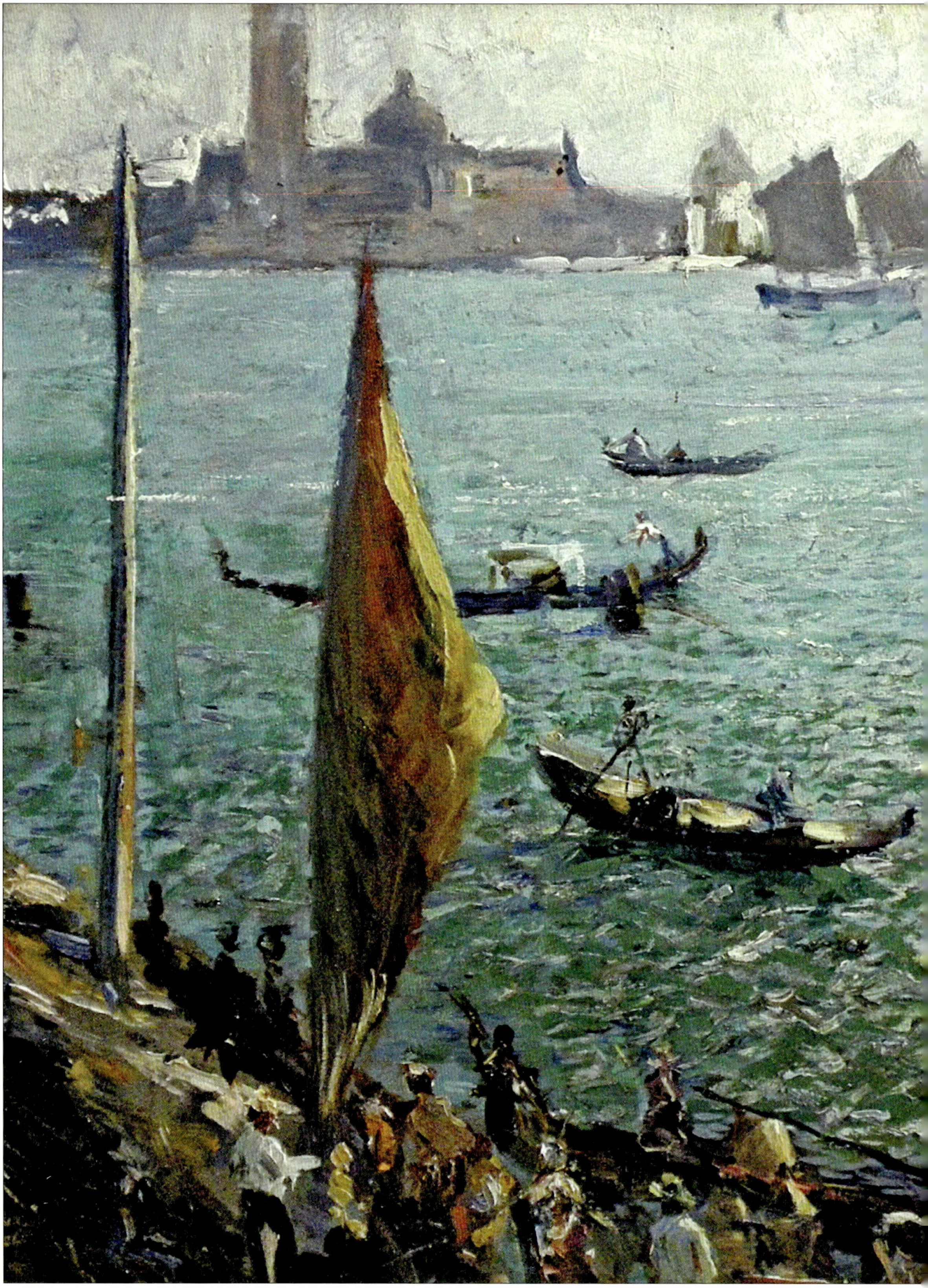

13. "Venezia, Riva degli Schiavoni". Oil on wood, 50 x 70 cm, dated and signed.

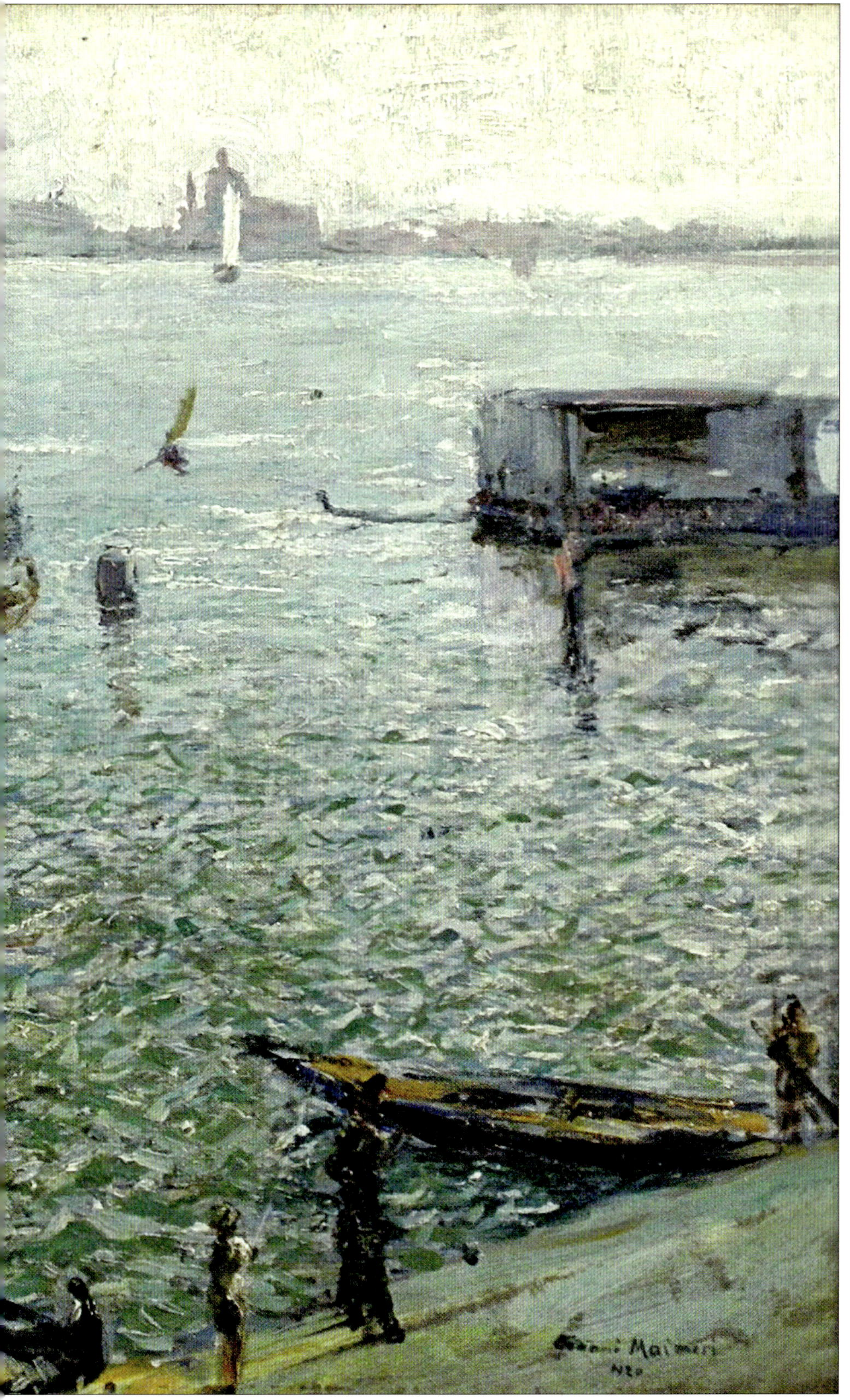

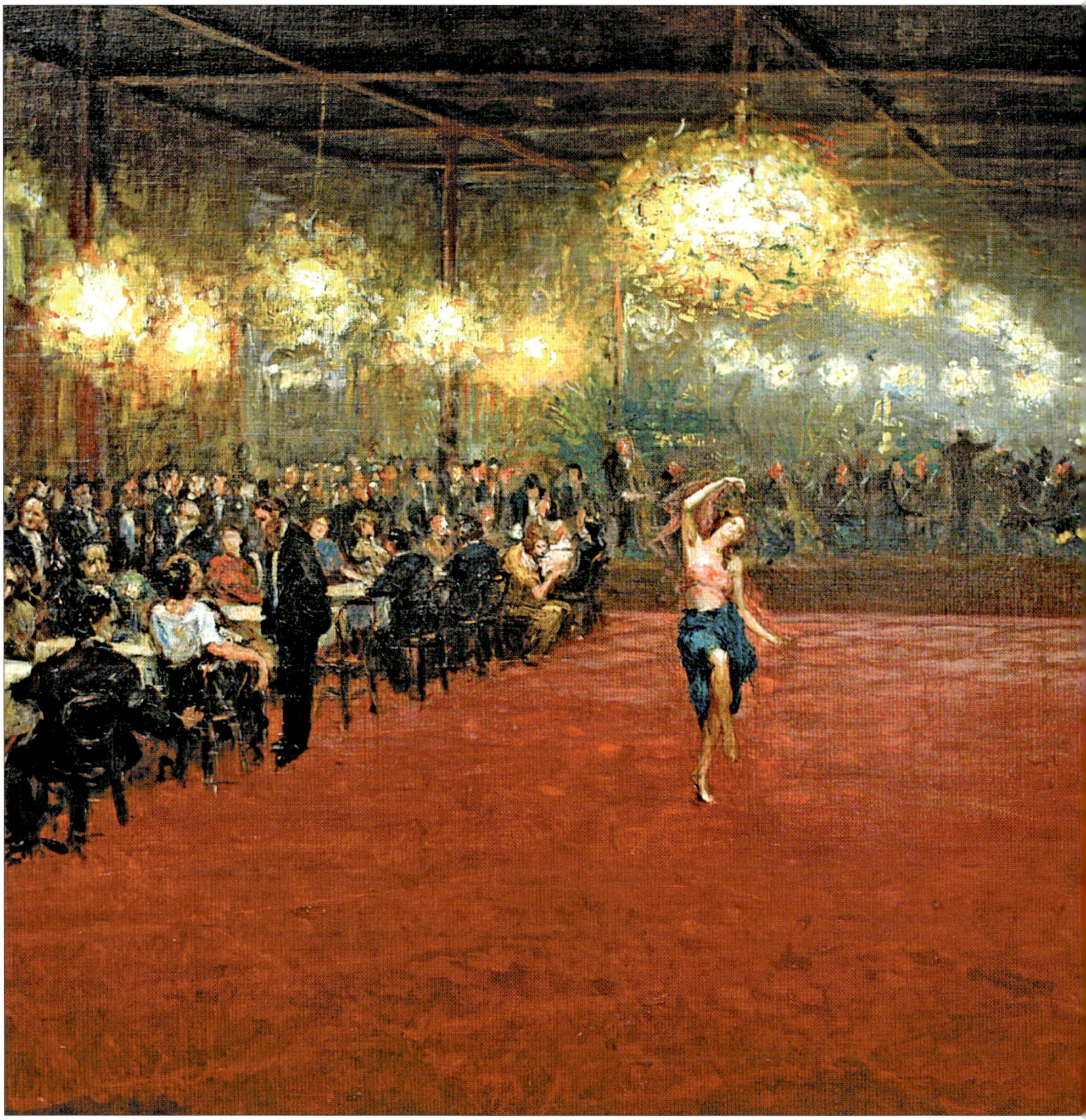

14. "Tabarin", 1914. Oil on canvas, 156 x 227 cm, dated and signed.

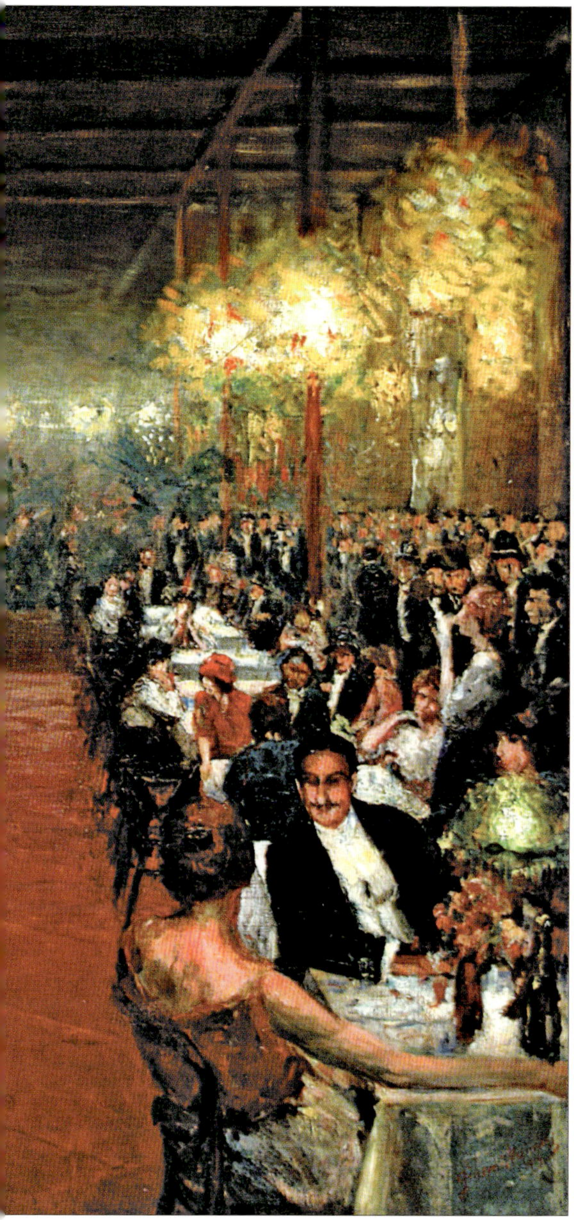

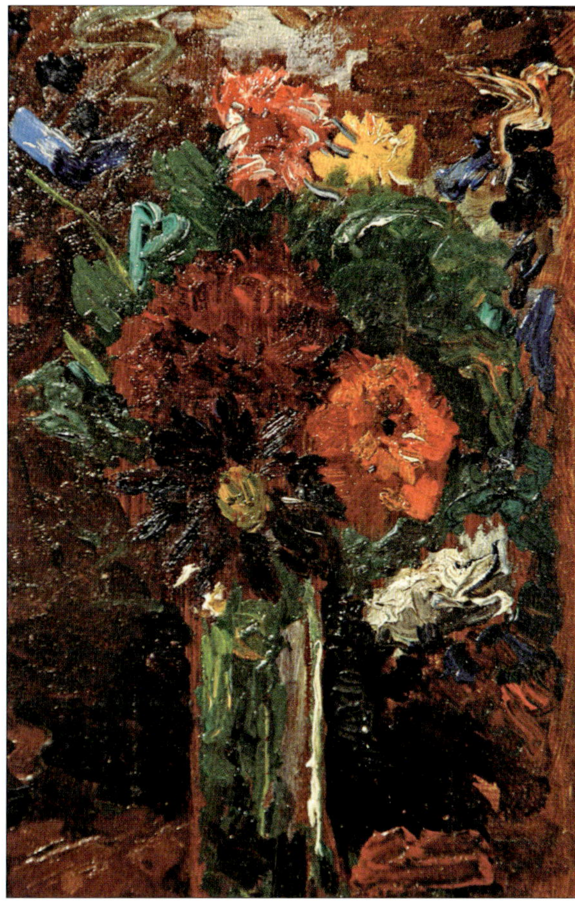

15. "Vaso di fiori", 1923 (First test with Maimeri colours). Oil on canvas, 17 x 11 cm, unsigned.

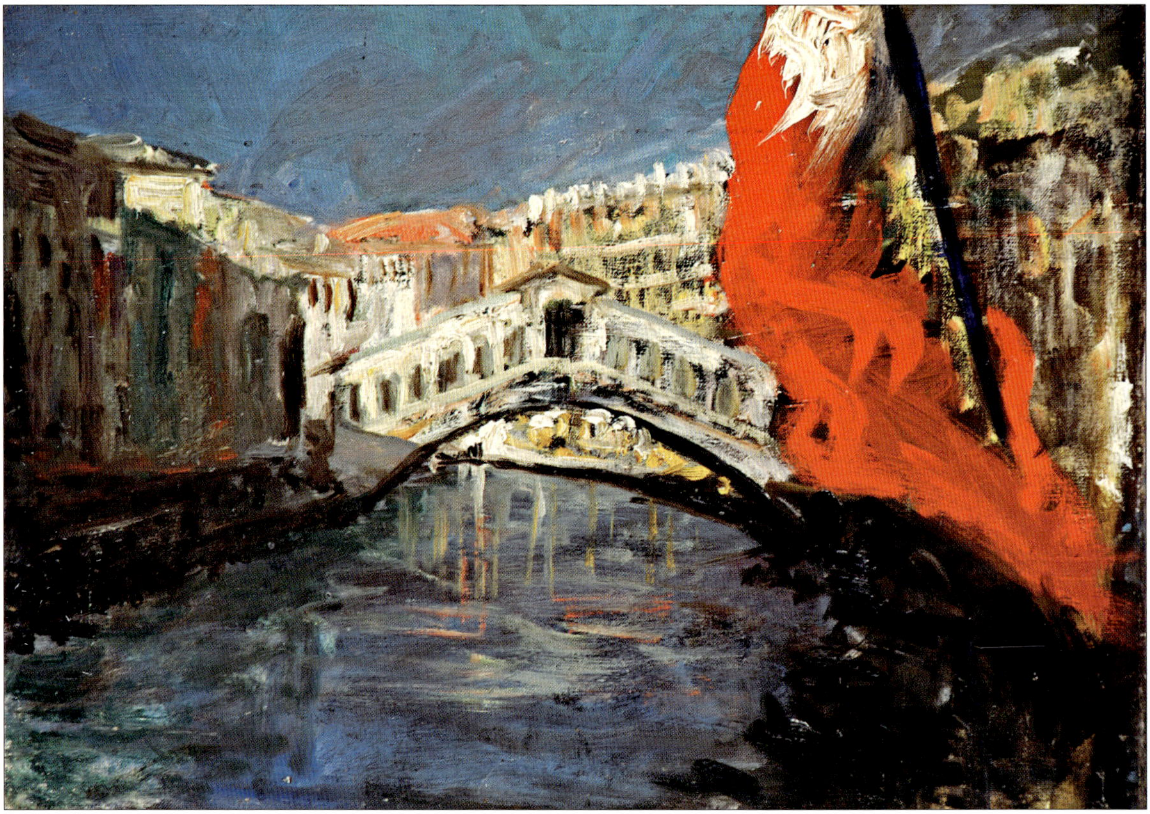

16. "Il Ponte di Rialto dalla finestra del comune", 1946. Oil on canvas, 40 x 54 cm, dated and unsigned.

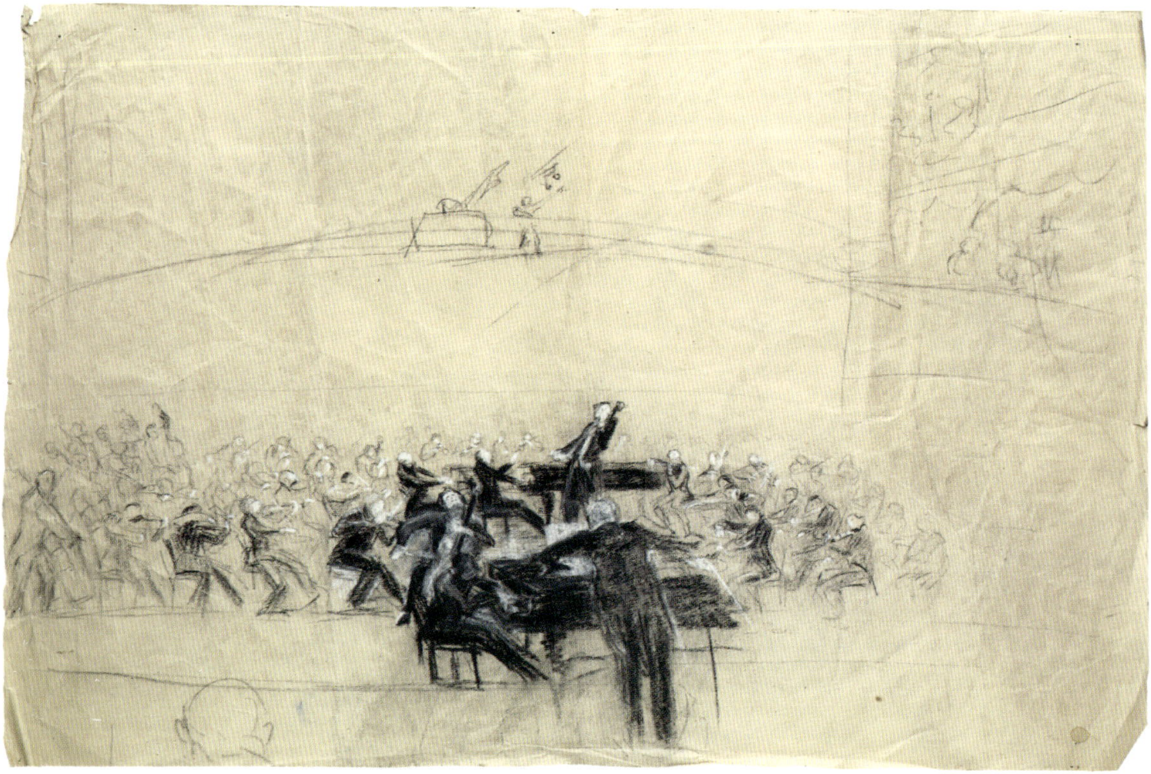

17. "Orchestra". Charcoal and pencil draw, 49 x 74 cm, undated and unsigned.

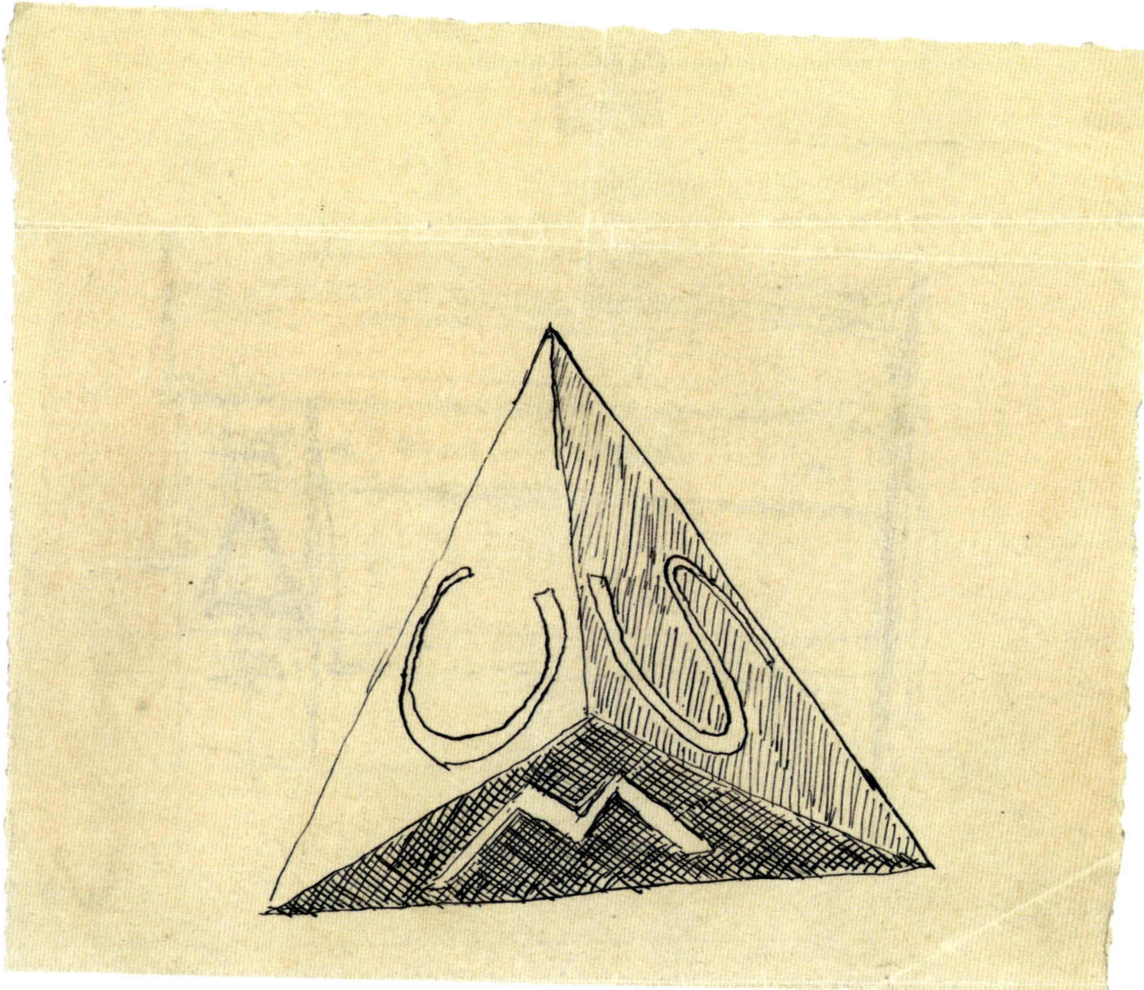

18. First Maimeri Trademark (Colori Stabili Maimeri), 1923.
Designed by Gianni Maimeri.

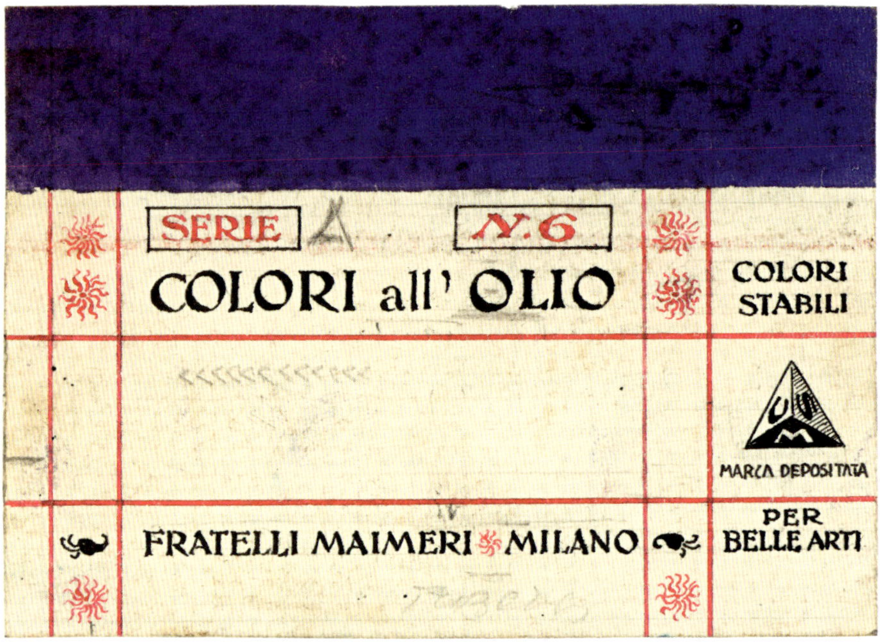

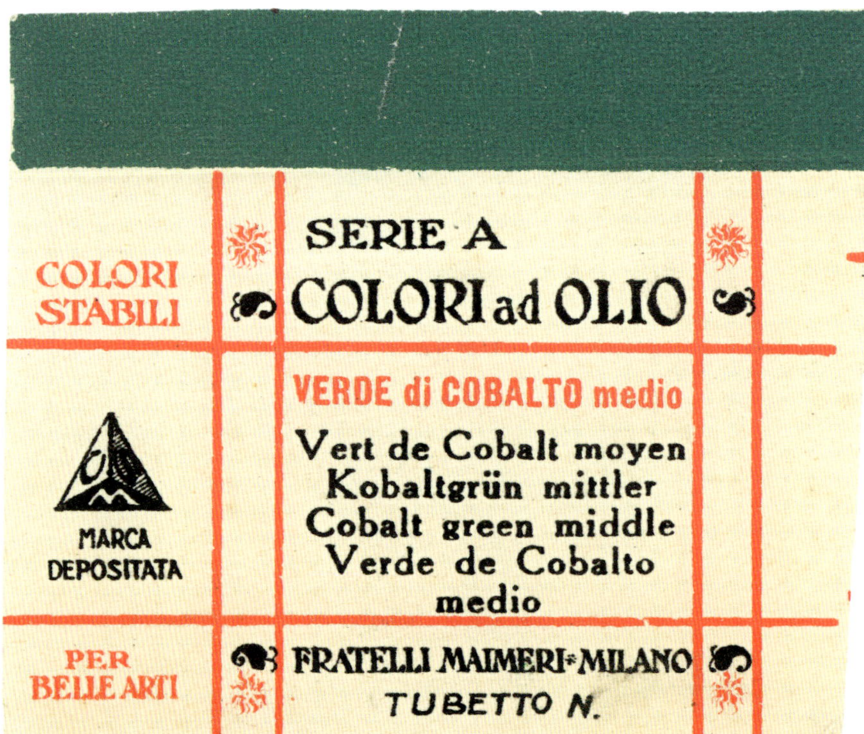

19-20. First Maimeri oil colours labels, 1923. Designed by Gianni Maimeri.

Appendix

Gianni Maimeri, biographical note

CARLO MIGLIAVACCA

1884

Giovanni (Gianni) Maimeri was born on 21 June in Varano (Varese) on Lake Comabbio, the second of three sons born to Leone Maimeri (1847-1912) and Zoe Bouffier (1851-1917). His father, an engineer, came from a family originally from Verona, was a Mason and socialist and at the time director of the big Borghi spinning mill in Varano, a little town in the Varese district for which he was town councillor. His mother was the daughter of Frenchman Adolphe Bouffier, a manufacturer of rolling stock (Usines Bouffier) and one of the founders of Elvetica (1846), one of Italy's first mechanics industries. His elder brother Antonio was born in 1883, his younger brother Carlo in 1886. Maimeri family life was spent between Varano, their place of residence, and Milan, where the boys attended school.

1890-1893

Attended the first three grades of elementary school in Varano.

1893-1899

Attended the fourth grade and five years of grammar school at the Collegio Calchi Taeggi in Milan.

1899-1903

Attended secondary school at the Cesare Beccaria Lyceum in Milan. Resided at a boarding house in Corso di Porta Romana right near the school.

1903

In June began to keep a journal in which until 1917 he wrote personal notes and above all aesthetic reflections regarding figurative art and music. The earliest pages are filled with caricatures of his family, sketches of interiors, landscapes and drapery studies, through which his approach to art took shape. His landscape studies were of the countryside around Varano and Lake Comabbio where he spent the summer months.

In July he told his parents of his intention to study art after getting his secondary school diploma, discarding his former decision to study architecture. In the fall he passed his make-up exams and on 10 November arrived in Venice, where he attended the school run by Vincenzo Rinaldo, a professor at the Accademia, in preparation for admission to the Venetian art school. He boarded with a family friend, attorney Adelchi Guaita, and made friends with his young son Carlo, later a renowned cellist.

1904

Early in the year, because of a quarrel, he left Rinaldo's school. After attempting to study under other Venetian artists, thanks to attorney Guaita's interest, he began to study two days a week with Giuseppe Vizzotto Alberti (Oderzo 1862 - Venice 1931) and together with a young colleague rented a room to use as a studio. He began to think of dispensing with further academic study, corroborated in this intention by his teacher, who considered him too old to follow a course of studies that in any case wouldn't teach him how to paint but only, and only partially, how to draw. In June he returned to Varano to take his military service exam.

Having finished secondary school, his brother Carlo enrolled in the Technische Hochschule in Zurich, one of Europe's most prestigious polytechnics, where he studied industrial chemistry.

1905

In January Maimeri returned to Venice. He visited the 6th International Art Exhibition, coming away with negative opinions of the Italian entries, in his view too interested in imitating famous foreign artists. During the summer he abandoned his studies, leaving Venice for Varano and Milan. In Leonardo Bazzaro (Milan 1853-1937) he found a "man *sans intrigues* or petty teachings" (*Journals*, p. 49) and a successor to Vizzotto Alberti as a mentor.

1906

He set up a studio in Porta Vigentina in Milan and began to frequent the city's cultural and artistic milieu; in particular, he formed ties with the Gruppo dell'Orologio (The Clock Group), a member of which was sculptor Antonio Carminati, who became a great friend.
He spent part of the spring in Varano, returning to the city to attend the inauguration of the National Exhibition held to celebrate the opening of the Simplon Tunnel. In the summer he spent a month with Bazzaro in Antronapiana in the Piedmont valley of Antrona above Villadossola. He began a portrait of cellist Carlo Guaita, son of Adelchi Guaita. In early December he departed for military service, which he performed in the 30th Infantry Regiment, billeted at the Sant'Ambrogio barracks in Milan.

1907

In the spring he did many landscape sketches inspired by the environs of Milan: the Naviglio (canal) at Corsico, the village of Chiaravalle with its abbey, the countryside beyond Porta Romana, the Lambro River at Moncucco.
On 20 July his brother Antonio, who had begun a brilliant career as mathematician, died of pulmonary tuberculosis, throwing Gianni into such deep prostration that he was exempted from his military service. He convalesced in Varano for the rest of the year, starting to paint again only in November. He finished his portrait of Carlo Guaita and began one of his paternal aunt, Bice, who lived with the family; at the same time he painted many sunrises and sunsets on wooden boards.

1908

In April his father resigned from the Borghi di Varano cotton mill, although he continued collaboration with it as consultant, and moved with the family to Milan, to a townhouse purchased at number 46, Via Boscovich.
On 14 June artist Emilio Gola made his first visit to the young painter's studio in Porta Vigentina, praised the colorist skills he saw in some of the paintings and noted a general lack of expression; he urged him to persevere, especially in figurative drawings.
Gianni spent the summer with his family in Courmayeur in the Aosta Valley, where he devoted himself to studying the Alpine landscape. He afterwards submitted his work in the mountains to Gola - by now his main reference, if not precisely his teacher - who considered them inadequate.
Brother Carlo was awarded a degree in chemical engineering from the Zurich Polytechnic.

1909

Gianni spent the winter and spring in Milan, studying the human figure, particularly that of an old beggar, and the urban scene. Fascinated by the artistic potential of the city's nightlife, he frequented Milan's coffeehouse concerts, filling a little album with drawings. After a lengthy study he painted "Bar Italia".
He spent the summer with his family at San Mamete Valsolda, on Lake Lugano, where he concentrated on comprehending the natural environment and the incidence of light on objects. Back again in the city, he recommenced seeing Gola in company with artist friends Carminati and Vittorio Castagneto, with whom he discussed the technical and psychological bases of being an artist. During the winter he painted his "Ritratto dell'Architetto Cesa-Bianchi". Art theory was interesting him more and more: he read Ugo Bernasconi's *Precetti e pensieri ai giovani pittori (Precepts and Thoughts for Young Artists)*, passing harsh judgment on it in the

detailed comments he wrote in his *Journals* (pp. 99-105). In the first half of the year his father had a studio built for him in the garden of their townhouse in Via Boscovich.

1910

Early in the year he met the girl who would seven years later become his wife: fifteen-year-old Anna Dossena (1895-1982), a model who posed for him and who returned his sentiments. This beautiful young woman, elegant and innately well bred, was destined to play an extremely important role in the artist's life and career. Thanks to a docile but extremely solid character, outstanding sensitivity and deep comprehension of human nature, she soon became indispensable in creating the conditions for her husband's serenity and happiness, supporting him when times were hard and encouraging him in his art as well as in his enterprises, even when they were nearly penniless. Her presence, always positive and reassuring, was characterised in particular by an ability to pursue a lifestyle imprinted with the rules of discretion and moderation, even in times of happiness and wealth. From this time on Anna Dossena would be his favorite model and continued to pose for him even after their marriage, her face appearing in many of his works. She was an authentic mainstay of the family she created with Gianni, and after his death promoted his work so that his name would be remembered.

Thanks to Anna, during this year he did a number of nude studies, at the same time painting under electric light a portrait of a lady friend of colleague Adone Comboni. He showed his work for the first time in exhibits at the Fine Arts and Permanent Exhibition Society in Milan ("Bar" [in pastel], "Lago di Antronapiana", "Signora seduta", "Bar", "Pezzente"). He spent his summer vacation with his family at Cunardo, in Valganna above Varese, where he was fascinated by the landscape's many caverns, later returning many times to the area.

1911

Between late 1910 and early 1911 he exhibited five works at the Famiglia Artistica di Milano ("Studi: Notte", "Studi: Venezia", "Studi: Strade", "Torrente", "Impressioni"). At the Exhibition of Independent Painters in Rome he presented his "Grotte di Cunardo". The paintings he sent in for selection by the annual Fine Arts and Permanent Exhibition Society show in Milan were rejected.

In early June he took a study trip to Rome and Florence, where he was particularly struck by the works of Raphael, Michelangelo, Titian, and classical statuary. He spent the rest of the summer vacationing with his family in Portovaltravaglia, on the Lombard side of Lake Maggiore. In the months to follow his father bought land in the area and began to build a house to his own design; it was finished a year later. According to his *Journals*, in Naples that fall he exhibited two paintings inspired by the Cunardo caverns and another of a mountain landscape ("Courmayeur").

Little more than a year after publication, in Milan, of the *Futurist Manifesto* (February 1910), he dealt very critically in his journal with the themes of this new movement. His theoretical thinking leaned more and more towards the function and potentiality of colour as he sought a way to go beyond what Gola taught him, good from the technical standpoint but passé from the aesthetic. In his journal are excerpts from Auguste Rodin's thoughts about art (A. Rodin, *L'Art*, Paris 1911), a monograph Elie Faure devoted to a highly admired artist, Eugène Carrière (E. Faure, *Eugène Carrière, peintre et lithographe*, Paris 1908) and the book on Impressionism by Camille Mauclaire (C. Mauclaire, *L'Impressionisme*, Paris 1904).

His brother Carlo finished a post-graduate course at the Technische Hochschule in Zurich.

1912

After a six-month illness his father died on 30 September. Gianni shouldered the burdens of finishing the house in Portovaltravaglia.

He attempted to display his work at the Venice Biennial but was refused. He spent the summer in

Portovaltravaglia and in December entered the first notes for an essay on colour in his journal. He visited Milanese museums in search of inspiration and a path for his artistic endeavours: at Castello Sforzesco he admired Van Dyck and Mosè Bianchi; at the Pinacoteca di Brera, Bernardino Luini, Cima da Conegliano and above all Veronese; at the Ambrosiana, Francesco Guardi.

One of his works was selected for the National Exposition held at Brera ("Racconto di fate"). In the catalogue it is listed as being by "Maimeri Gianni da Varano". With "Nudo di donna - Interno" he participated in the Fondazione Fumagalli Competition held at the Accademia di Brera on the theme of the female figure "from life". Between the end of the year and the start of the next he showed a number of paintings at the Famiglia Artistica ("Ricordo di Venezia", "Crepuscolo", "Pomeriggio", "Ottobre", "Lo stagno").

1913

He participated in the Fondazione Canonica competition on "women at their toilettes" held at the Brera. The work he entered, probably "La donna e lo specchio", did not win a prize but was chosen as one of the three finalists by celebrated portraitist Cesare Tallone, painting professor at the fine arts academy. The painting was later purchased by the Italian royal family.
Maimeri's journals increasingly recorded negative judgments on Futurism and Cubism. He considered the two movements devoid of any real innovation and bearers of a false idea of liberation from 19th century painting traditions. He instead admired the work of Eugène Carrière, immune to Impressionism and a reference point for a "new tradition" (*Journals*, p. 155).

1914
In the spring, after the Venice Biennial committee turned down another painting, Gola went to see Maimeri, upset by this injustice. The painting, "Mezzanotte", had got favourable votes from Gaetano Previati, Adolfo Wildt and Baldassare Longoni. Maimeri attributed the refusal to the effects of a circular he had written in favor of making Gola a member of the jury for the international exposition. The candidate preferred to him was, instead, Carlo Cressini, backed by Famiglia Artistica di Milano. During this period he painted the large canvas, "Tabarin", new fruit of his observations of city nightlife. He exhibited "Andante Sostenuto" and "Raggi e languore" at the Annual Exhibition of the Società Permanente.
His brother Carlo found employment in Milan at the Ledoga (Lepetit) chemical laboratories.

1915
He exhibited "Mezzanotte" and a pastel portrait at the Annual Exhibition of the Società Permanente.

1916
He showed his works twice at the Permanente: in the context of the annual exhibit (with a diptych, "Notte - scenette") and at the Fine Arts Exhibition ("Studio", "Alla lucerna", "La casa abbandonata"). He also exhibited at the Brera National Exhibition with a pastel, "Il sofà".
When Italy entered the war the draft board in the Varese district, of which Varano Borghi was part, gave him exemption status because he had dislocated his left shoulder. In the fall his mother fell ill and was taken to Portovaltravaglia where Gianni lovingly attended her. Anna wrote to him from Milan, keeping him updated on the singing lessons he had encouraged her to take.

1917
On 24 April his mother died in Portovaltravaglia. Gianni remained there for some time to settle some family business and have the estate's vines grafted. In the fall he married Anna Dossena. He was again exempted from military service because of problems with his right lung.

1918

From 18 May to 2 June he exhibited seventy-four works at the Galleria Geri in Milan. This was his first one-man show and consisted of figurative works, night scenes of Milan, views of Venice and landscapes, primarily wintertime. Raffaello Giolli wrote the catalogue; reviews in Milanese newspapers were good, dwelling mostly on his explorations into light and colour.

Brother Carlo married Josette Biraghi Lossetti, whom he had met during the war when serving as an officer at the Piedmont Military Chemicals Center in Pieve Vergonte, near Vogogna where the Biraghi Lossetti family owned a big villa. From this marriage would come three children: Antonio (1919-2003), Gianni (1921-1949) and Ester, born in 1928.

1920

He appears to have joined the Cooperativa Artisti per Materiale Belle Arti (Artists Cooperative for Fine Arts Materials), a joint-stock cooperative founded that same year. At the annual exhibition of the Società Permanente he showed the painting "Giardino".

1921

The rooms at the Palazzo della Permanente welcomed him among participants at the Exhibition of Veneto, Tuscan and Lombard Artists, where he showed "Nel bosco", "Bacino di San Marco", "Riva degli Schiavoni".

He planned a trip, never undertaken, to Kenya (then a British colony) following the Catholic mission to Nyeri. This must have been a painting trip, since he had an agreement with Milanese gallery owner Lino Pesaro to show the work done during the trip during the 1923-1924 exhibition season. He worked for six months in Gola's studio.

1922

His eldest daughter, Zoe (1922-2004) was born. At the Brera National Exhibition he showed "Nel giardino"; in the catalogue he was still listed as "Maimeri Gianni di Varano".

1923

Maimeri was beginning to develop the idea, shared with brother Carlo, of setting up an artists' paints factory. The two studied what was imported from abroad to check the feasibility of the enterprise and drafted an initial estimation of product variety and the machinery needed. In their Portovaltravaglia home, following his brother's technical instructions, Gianni engaged in his first experiments with mixing pigments and made his first assessments of paint tube type and volume, sending the results to Carlo in Milan. In the meantime he was also handling the sale of the house, which his father had designed, in order to finance the nascent business; nor did he neglect to paint.

After months of research into mineral pigments and drying oils, in September the two brothers drafted a definitive design for the layout of the new factory, including expected outlays and income, and fruitlessly submitted it to chemical giant Max Meyer, already making industrial paints. At the time Carlo was involved, on behalf of Ledoga, in designing and outfitting a chemicals plant at Darfo, in Val Camonica (Brescia).

Over the winter Gianni looked for an industrial building suitable to housing the business as well as his family (the brothers sold the Via Boscovich townhouse, too). In December attention focused on a former oilseed pressing plant at Barona, on the southwest outskirts of Milan, which became the company's first premises.

1924

Work started in January to fix up the production plant and apartment at Barona and continued through spring; by the end of the year the machinery was installed and running. Gianni continued his research into colour production here, while from Darfo Carlo sent him cost estimates for paint tubes, labels and cardboard boxes. Two partners appeared: Trentino baron Silvio A Prato, a chemist, friend and co-student of Carlo's at the Zurich Polytechnic, and Giovanni Ticozzi, Ledoga manager in Darfo and earlier employed by engineer Leone Maimeri in Varano. During the summer production of the first industrial samples began, albeit slowly, since plant equipment was not yet complete. Gianni was tasked with plant management.

1925
With the colour range decided on, free samples were sent to artists, dealers and representatives throughout Italy, accompanied by a technical description of the new methods used to prepare the pigment and oil mixtures. Appreciative thanks for the samples and admiration of production quality were quick to come.
June 27 saw constitution of the Società Anonima Fratelli Maimeri S.A., which replaced the previous Fratelli Maimeri, put into liquidation. The partnership agreement was signed by Carlo and Gianni Maimeri, Silvio A Prato and Giovanni Ticozzi. Company capital was ten thousand lire, divided into forty shares worth 250 lire each (Gianni Maimeri invested 3,250 lire, Silvio Prato 3,000 lire, Carlo Maimeri 2,750 lire and Giovanni Ticozzi 1,000 lire). Gianni was appointed president and Carlo, who had meantime left Ledoga, managing director. In July a report was written detailing the aims and financial status of the company, to be submitted to any potential new financers. There was little money in the till, due to the great amounts spent to start production, but Prato often saw to its replenishment, permitting the company to pay its most pressing bills.

1926
In January the brothers went to Venice to seek a distributor for Maimeri products. They put their case before the Sindacato Nazionale delle Arti del Disegno (Italian Artists and Art Supplies Union), requesting protection against imported products and promotion for their colours. In exchange they declared themselves willing to give the union control over production quality and offered a 25% discount to artists belonging to the organization. But negotiations did not lead to anything concrete. They contacted Max Mayer again, in the hope of getting a new partner able to invest new capital, but here too without success.
More and more letters of praise were arriving from artists who had been sent free samples of the colours. Among these were Arturo Tosi and Leonardo Bazzaro in Milan, Giuseppe Fraschetti in Florence, Umberto Coromaldi (professor at the Regia Accademia di Belle Arti in Rome), Guido Trentini (professor at the Accademia di Verona), and Vincenzo Irolli in Naples.
In a letter sent to Carlo on 12 August, Gianni said he was painting "at a breathless rate" despite the burdens of managing the plant (ApfM, *Ditta* n. 3). Later in the month he took a vacation on the Lido in Venice.
By the end of the year the company's financial situation was still critical and the partners decided to appeal to the Economy Ministry to levy heavy customs duties on imported artists' colours. Praise for the products continued to arrive from artists, and in December Gianni got a letter full of compliments from Pompeo Mariani in Bordighera, who to repay the gift of two boxes of colours sent him a "sea sketch". Also in December, company capital was raised to 100,000 lire: the Maimeri brothers' new shares were paid for in advance by baron A Prato. To thank his friend for the loan, Gianni sent him a painting; he also did restoration work for the baron. Son Leone was born.

1927
In April the brothers turned to Cipriano Efisio Oppo, secretary-general of the Artists and Art Supplies Union, urging him to protect Italian products from foreign competition. They invited the artist to visit the plant, as they did Margherita Sarfatti, the art critic at the time responsible for the success of the Novecento Italiano movement. They sent her boxes of colours to donate, at her discretion, to needy artists.
In June Gianni went to Rome for a personal meeting with Oppo, from whom he obtained general support for union help in running the company. Meanwhile, and despite a considerable increase in sales, the cash-flow situation halted production for a time, limiting the company to selling off the material it had in stock. Disheartened by the crisis in

the company, Gianni spoke of his discouragement in a letter to friend Adelchi Guaita in Venice, himself involved in promoting Maimeri products in his city: "Having devoured so many numbers over these years, I who couldn't abide them, now to risk finding my hands full of the number by itself worth less would be a fine result" (letter of July 6, ApfM, *Ditta* n. 3).

1928

At the beginning of the year, persisting financial difficulties led to an initial cooling of relations between the Maimeri brothers and Prato. He asked the two to step down from managing the company and entrust its running to others. Subsequent mediation had the brothers resign from their positions with the company and take up others, as external technical-artistic and chemistry consultants, with consequent salary reductions. The search for new partners continued. Gianni spent the month of August at the Lido in Venice, returning in September "empty of pocket" but "full of paintings" (letter written to Carlo on September 2, ApfM, *Ditta*, n. 4). The baron clearly stated his wish to cede his shareholdings and Gianni contacted the company's Tuscan representative, Sante Faccini, asking if he might be interested in buying them.

1929

As suggested by Carlo Accetti, president of Famiglia Artistica di Milano, Maimeri requested admission to the historic society as a "member painter". His application was accepted on 4 February but by 7 November he had already resigned, disappointed at his works' exclusion from the society's annual exhibition. Even one of the works he submitted to the Art Exhibition of the Fascist Regional Union of the Fine Arts of Lombardy was refused ("Ombra e sole - Il Mulino") while he managed to exhibit "Afa sulla laguna", purchased by the Province of Milan. He participated in the Third Landscape Exhibition in Bologna, winning the gold medal awarded by the Communications Ministry for "Fine d'inverno". Between May and June he had two works ("Via Senato" and "Via Fatebenefratelli") on display at the Canals of Milan Exhibition held in the Palazzo della Permanente to sadly salute the covering of the canals in the heart of the city, sacrificed to pave the streets for motor vehicles. He donated two works to the collection of the Landscape Museum in Pallanza, on Lake Maggiore.
In June the company was struck a hard blow by the bankruptcy and desertion of accountant Angelo Pietrasanta, board member of Fratelli Maimeri S.A. and an influential procurer of credit for the company. Gianni and Sante Faccini (who had become agent for almost all of central and southern Italy) continued their attempts to find a buyer for Silvio Prato's shares. July saw foundation of Fratelli Maimeri & C., destined to take over the joint-stock company about to go into liquidation.

1930

He was invited to display his work at the inaugural exhibit for the Galleria Amadei in Milan (22 February - 15 March) where he showed "Fine d'inverno" and "Fiori". In March he mounted a big one-man show at the salon held by the *Giornale dell'Arte*, also in Milan. For the occasion he selected forty pieces: figures, nocturnes, still lifes, views of the Milanese canals, snowfalls, views of the Lido di Venezia, depictions of the Barona and Valtravaglia landscapes. In April the executive committee of the Italian Landscape and Picturesque Monument Association charged him with directing its newly-forming chapter for Milan and province. In May he was made a member of the Fine Arts and Permanent Exhibition Society of Milan. From 1 to 30 June he exhibited fifty-one works at the Bottega d'Arte Gallery in Livorno: among them were many of the outskirts of Milan, inspired by the environs of the Barona paint plant. Old friend Fernando Viviani wrote the preface to the catalogue. The jury for the 17th International Art Exhibition in Venice turned down two of his paintings: "Il Lambretto d'autunno" and "La vetreria Lucchini a Portovaltravaglia". Instead, he succeeded in presenting four works in Milan at the show of watercolours, pastels and works in black and white held

at the Palazzo della Permanente: two lithographs, "Il Naviglio di Via Olocati" and "Il Naviglio di Via Senato", and two pastels, "Pezzente" and "Lo scialle".
From October to November, on behalf of the Milanese chapter of the Italian Landscape and Picturesque Monument Association, he wrote a huge document containing the association's comments on, proposals for and amendments to the broadened urban renewal plan prepared by the city government. The document contained a detailed list of twenty-two urban areas within the historical center of town whose characteristics needed to be retained because of their historical importance and to preserve the cityscape. The "battle" was picked up by the press, asked to give visibility to Maimeri's and the association's views. The issue was given special coverage by *Il Giornale dell'Arte*, which from September through November hosted several of his articles. To back opposition to the urban renewal plan he also appealed to architect Luca Beltrami, celebrated restorer of Castello Sforzesco in Milan but long a resident of Rome, considering him the most famous defender of the city's original appearance. Beltrami replied that he was not acquainted with the issue of the new plan, but would send documents relating to his opposition to the previous urban renewal plan. His observations were welcomed by the Higher Fine Arts Board.
The first months of the year saw start-up of the procedures to liquidate Fratelli Maimeri S.A., accompanied by intense correspondence between Gianni and baron Prato concerning the latter's heavy outlays to the company and the two Maimeri brothers. Daughter Zoe, a promising pianist, was admitted to the Milan Conservatory.

1931

Three works were submitted to the First National Art Quadrennial in Rome, but without success. In any case, he visited the exhibition, held from January through June. The same fate befell three paintings sent for judgment by the committee organising a show for the Society for Promotion of the Fine Arts in Turin. However, between April and May, at the members' show organized by the Fine Arts and Permanent Exhibition Society in Milan, two works of his were shown ("Meriggio" and "Risaia") and at the same time he exhibited with Giovanni Lentini at the Galleria Pesaro in Milan; eighty-three of his works were on display here, consisting of figures, landscapes, nocturnes and views of the canals. The Province of Milan bought the painting "La coppa" at this show. Between June and October he was one of the artists exhibiting at the 84th Members' Exhibition held by the Florence Society of Fine Arts, to which he belonged. At the exhibition, held at the Palazzo delle Esposizioni in Montecatini Terme, his "Uccellini" was on display.
While Fratelli Maimeri & C. was having a hard time staying in business, the complex liquidation of Fratelli Maimeri S.A. proceeded. To meet the due-date for a debt incurred with the Calderari & Moggioli bank in Trento, Gianni had to offer sixty-odd paintings in guarantee of payment; for the same reason Carlo had to mortgage his house at Via Machiavelli 32, bought in his wife's name between 1921 and 1922. Gianni also committed himself to definitively liquidating his debts to Silvio Prato, and put up as collateral a second lot of forty paintings. The transaction was concluded in March, and from that time on there were no further relations, of business or of friendship, with Prato. In the meantime, new partner Sante Faccini was hectically traveling the peninsula trying to get orders for the company, still in troubled waters.
Gianni's opposition to the urban renewal plan for Milan continued; the city project, partially approved in November 1930, did undergo some changes, but these were not sufficient for him and in January 1931, again in the name of the National Landscape and Picturesque Monument Association, he presented an official petition to city officials appealing for a halt to work, but this had no effect.

1932

In January, together with Leonardo Bazzaro, Galileo Chini and Alessandro Mazzuccotelli he showed his work at a group show held at the Galleria Pesaro. With "Tramonto (Cascina alla Barona)" he participated in February and March at the 3rd Art Exhibition of the Fascist Regional Union of the Fine Arts of Lombardy Brera Biennial organized at the Permanente. In the following months he had four pictures on view at the Permanente members' show ("Strada campestre", "I due canali. Nevicata", "Ruscello tra la neve") and from July to October he showed two works (a landscape in oil and a figure in pastel) at the 2nd Montecatini Spa Art Exhibition sponsored by the Florence Society of Fine Arts. Two paintings submitted for selection by the committee for the third regional Lombard show organized by the Fascist Regional Union of the Fine Arts were refused. Between November and December he had work on display at the Self-Portrait Exhibition organized by the Fine Arts and Permanent Exhibition Society in Milan. At the end of the year he wrote the first pages of a long philosophical treatise on colour (still unpublished) provisionally titled *Trattazione di uno studio sul colore* (Treatise on a Study of Colour), to which he would often return in the years to follow. In Milan he happened to drop in at an auction where up for bidding were two of the paintings he had offered as guarantees for liquidating Fratelli Maimeri S.A. and for repaying his debt to Silvio Prato ("Il pianoforte" and "Nevicata"). Finding them priced well below the expert's appraisal, he challenged the sale and indignantly reminded the liquidator of the agreements stipulated when the works were ceded in guarantee.

1933

He tried to enter the IV Regional Lombard Exhibit organized by the Fascist Regional Union of the Fine Arts but the painting "Interno illuminato" is rejected. In the spring he presented two paintings ("Mezzogiorno" and "Estate") for admission to the 1st Inter-regional Exhibit being organized by the Florence chapter of the Fascist Fine Arts Union, with both refused. Piqued by this further exclusion, he wrote to painter friend Donato Frisia, a member of the acceptance committee, asking him for explanations. Frisia replied that he had backed him but found himself in a minority, and advised him to appeal to the Union. Gianni declined the suggestion and sent him a memo about what he called the "Maimeri Affair", or the chronicle of the many (to him suspicious) exclusions of his works from union shows (ApfM, letters written on April 1, 4 and 5, 1933). Between October and November he had three paintings in the Exhibition of Modern Italian Painters organized by the Nuova Vita cultural association in Milan; other exhibitors were Aldo Carpi, Guido Tallone, Donato Frisia, Anselmo Bucci and Leonardo Dudreville.

He continued to manage the business, staying in constant contact with Sante Faccini who, albeit with effort, was able to keep orders coming in from all over Italy. This partner and traveling salesman also managed to find him some special jobs, touting his artistic abilities to several collectors who intended to "divide" works in their possession in order to get two paintings from canvases painted on both sides. He advised his brother Carlo, who had already expressed this intention, to definitively leave the company and profitably use his skills as a chemist in another field. After a number of years working as a consultant for various companies and the Milan Polytechnic, Carlo found a job with the ACNA company (Montecatini), setting up and directing chemical research laboratories in various parts of Italy. Carlo's first posting was with Bussi Officine, near Sulmona, to which he moved with his family.

1934

Gianni took part in the Permanente's spring members' show with "Rose thea", "Carpa", "Fiori di Tagete", "Crepuscolo sul Lambro", then with "Inverno - Paesaggio lombardo" in the 4th Landscape Art Exhibition held in Bologna, curated by the Italian Landscape and Picturesque Monument Association, and afterwards at the 5th Lombard Interprovincial Fascist Union of the Fine Arts Exhibition with "Fiori". He sent the "Ritratto del Generale Chinotto" and a war episode from the general's career to the War and Victory Competition in Rome, sponsored by Queen Elena. In No-

vember he was among the painters exhibiting at the First General Exhibition of 21 Contemporary Artists held at the Galleria Vitelli in Genoa: his five contributions were "Inverno lombardo", "Ruscello", "La bottiglia", "La statuina" and "Fiori". To bring in some money for his family he also did minor restoration jobs.
As a member of the company, Sante Faccini wanted Carlo to define his own position by investing in it but Carlo, who had just begun another line of work, did not intend to risk any more money on the business. Gianni urged him to free himself of any financial commitment to Fratelli Maimeri & C. although Carlo was never afterwards remiss in giving him technical advice and moral support. At this time Carlo and his family were in Cengio (Savona) on behalf of Montecatini, while waiting to start up a new laboratory in Cesano Maderno near Milan.

1935
He sent the Permanente "I due canali", "Lavandaie al ruscello", and "Carte e chicchere sul tavolino" for its members' show. This last painting was bought and donated to the City of Milan by the Legato Sallustio Fornara, a fund set up to promote Lombard artists (the painting is now in Milan's Civica Galleria d'Arte Moderna). In May he again exhibited at the Permanente (this time with "Lo stagno dei pesci rossi") for the Lombard Interprovincial Fascist Union of the Fine Arts Exhibition. The work he submitted for admission to the 2nd National Quadrennial in Rome was refused. In March his painting depicting General Chinotto's heroism was given an award at the Rome competition. He participated in the winter session of the same exhibition with a painting of "Il Po a Monticelli".
The government passed a total ban on importing packaged artists' colours and put a heavy customs duty on the raw materials, something that gave Maimeri & C. an economic boost and permitted it to expand.
Daughter Zoe was attending her seventh year of piano lessons at the Milan Conservatory; in this same year she also graduated with top grades from the Conservatorio di Santa Cecilia in Rome.

1936
Between April and May his "Strada rustica" and "Fiori di primavera" were on view at the spring members' show at the Permanente, and he afterwards exhibited at the 7th Milan Interprovincial Fascist Fine Arts Union show; there, the city's *podestà* bought for the Milan Museum the only painting he had on view, the oil "Neve" (now in the Milan Civic Gallery of Modern Art). In November he gave the Permanente his painting "Budda" so it could be sold to raise funds for the society. The company was also beginning to make tempera paints, much in demand at that time. Fratelli Maimeri & C. left its original premises at the Barona (Mulino Blondel) and relocated not far away in Via Ettore Ponti, along the Carlesca canal.

1937
He exhibited two paintings at the Permanente's spring members' show: "Astri Selvatici" and "La chiesa della Barona". At the autumn edition of the same show he exhibited "Sotto il fogliame". He sent another work, "Cassoncino e rose di maggio" to the Italian Art Show organized at the Universal Exhibition in Paris.

1938
In April and May he had three paintings at the spring Permanente exhibition: "Studio del pittore", "Il piccolo pittore" and "Neve nel bosco". In July the Milan Hospital Board commissioned him to paint a portrait of attorney Romolo Quaglino, a deceased benefactor of the Ospedale Maggiore, for the board's portrait collection. Between November and December he exhibited at the Permanente's autumn members' show with "Visciole", "Il giardino Melzi a Porta Romana" and "Ciliegie". Fire destroyed part of the company's warehouse. Gianni and his family moved to an apartment in Corso di Porta Romana (at the time named Corso Roma).

1939

In June he had a one-man show at the Galleria Gian Ferrari in Milan, with forty works: landscapes, interiors and still lifes. In October he got a new commission from the Milan Hospital Board, this time for a portrait of benefactress Maria Biraghi for the board's collection. A month later he exhibited three works ("Mezzogiorno", "Cassoncino e rose di maggio", "Paesaggio lombardo") at the Twelve Contemporary Artists show held in Milan at the Galleria Nova.

He and his family left the Corso di Porta Romana apartment, moving to Corso Italia 40. Carlo was made director of the Rumianca company's research laboratory at its plant in Pieve-Vergonte, where he had been stationed during the war.

1941

Gianni tried, in vain, to be included in the 3rd National Fascist Fine Arts Union Exhibition. Instead he had another one-man show at the Galleria Gian Ferrari from October to November, displaying some fifty-one paintings, mostly landscapes and still-lifes, and with a catalogue introduction by Gustavo Botta. In March his daughter Zoe gave her first public concert at the Milan Conservatory.

1942

Wartime restrictions obliged him to seek permission from the Milan Defense Command to paint the Po River at Spessa (Pavia), the Adda at Rivolta d'Adda and the Ticino at Bereguardo. Permission was granted, with the specific prohibition to paint military or industrial installations, including bridges.

1945

During the bombardments on 31 March, part of the Fratelli Maimeri & C. plant in Via Ettore Ponti 10 was destroyed. The damage report written in June put the total sum at 2,700,000 lire.

1946

On 30 July there was a fatal accident at the Maimeri plant. Employee Maria Bonasegale, with the company since its foundation, lost her life from burns incurred during a fire in the room where she was working on a paint.

1947

In a letter to his cousin Gustavo Botta, Gianni said that he was neglecting his painting but had finally put the company's business in order and was now being helped by son Leone (ApfM, *Ditta* n. 5). Sante Faccini decided to move back to Florence and his shares in the company were bought by a friend of Gianni's, Paolo Pini of Rome, with the help of a Como company that made oils. Carlo was hired by the Carlo Erba pharmaceuticals company to direct its research laboratory.

1948

The Ariel publishing house in Milan put out the volume *Gianni Maimeri. Tre saggi critici*, a collection of three critiques of his works written in 1931, 1939 and 1941 by his cousin Gustavo Botta, who died that year.

1948-1949

From December to January he held a broad-ranging one-man show at the Galleria Salvetti in Milan. The show opened with his "Autoritratto" of 1930 and continued with a number of nocturnes, twenty compositions and still lifes, twenty-two landscapes and a good thirty views of the defunct Navigli, painted at dawn and at sunset between February and March, 1929. A large number of drawings from life of soloists, musicians and conductors, from Busoni to Toscanini, completed the show, whose catalogue introduction was written by Gustavo Predaval. Seriously troubled family circumstances obliged Maimeri to sell the paintings of the canals, which were bought by the Ramazzotti company (and are now dispersed in private collections). The artist put a condition on selling the series, with a specific buyer's clause in the contract: "We undertake to keep the series of thirty paintings intact in order to maintain the character and worth of the artistic inspiration you demonstrated in wishing to depict many aspects of the Milan canals when their demise was decreed, for posterity and in filial tribute to love for the city, for art and for one of the picturesque features of our Milan of yesteryear."

1949

In June he had another show at the Gian Ferrari gallery and wrote the final pages of his *Trattazione di uno studio sul colore* (Treatise on the Study of Colour), begun in 1932. The philosophical slant of the work was counterbalanced in the postwar years by his *Trattato della pittura* (Treatise on Paint) in which he left to posterity acute and documented thinking about the technical aspects of painting, gained through so many years of artistic and industrial experience.

He spent the summer in Rome and on Ischia, painting many views of both the city and the island. In November he held a show of his works at the Galleria La Barcaccia.

1950

From 1 to 15 February he exhibited a series of Ischia landscapes and views of Rome at the Galleria Travaglini in Milan.

1951

Maimeri died on the afternoon of 25 November at the age of sixty-seven.

Catalogue of Exhibitions of the Works of Gianni Maimeri (1910-2008)

MAITE ROSSI AND PAOLA TRAVAGLIO

We present here, in chronological order, a list of exhibitions of the paintings and drawings of Gianni Maimeri, of use for reconstructing his artistic career and fortune among the critics.

This catalogue is based on copies of the press kit, exhibition catalogues and publications stored in the Maimeri family's archive. This work was completed by reading labels, inscriptions and stamps applied to the backs of the works by the artist, providing information of use for reconstructing the various corpuses of works exhibited in individual exhibitions.

Exhibition of the Fine Arts and Permanent Exhibition Society, Milan, Palazzo della Permanente, 1910.

Works exhibited:
"Bar" (pastel), "Lago di Antronapiana", "Signora seduta", "Bar", "Pezzente".

Group Exhibition, Famiglia Artistica, Milan, December 1910 - January 1911.

Works exhibited:
"Studi - Notte", "Studi - Venezia", "Studi - Strade", "Torrente", "Impressioni".

Exhibition of independent painters, Rome, 1911.

Work exhibited:
"Grotte di Cunardo".

Exhibition of independent painters, Naples, 1911.

Works exhibited:
"Grotte Cunardo (2)", "Courmayeur".

Brera National Exhibition, Milan, 1912.

Work exhibited:
"Racconto di fate".

Fondazione Fumagalli Competition, held by Accademia di Brera, Milan, 1912.

Work exhibited:
"Nudo di donna - Interno".

Famiglia Artistica, Milan, 1912.

Works exhibited:
"Ricordo di Venezia", "Crepuscolo", "Pomeriggio", "Ottobre", "Lo stagno", "La donna e lo specchio" (?).

Fondazione Canonica Competition held by Accademia di Brera, Milan, 1913.

Work exhibited:
"La donna e lo specchio" (the work did not win a prize, but was one of three finalists).

Annual Exhibition of the Società Permanente, Milan, Palazzo della Permanente, 1914.

Works exhibited:
"Andante Sostenuto", "Raggi e languore".

Annual Exhibition of the Società Permanente, Milan, Palazzo della Permanente, 1915.

Works exhibited:
"Mezzanotte", "Ritratto di mia moglie" (pastel).

Annual Exhibition of the Società Permanente, Milan, Palazzo della Permanente, 1916.

Works exhibited:
"Notte - scenette" - diptych.

Fine Arts Exhibition of the Società Permanente, Milan, Palazzo della Permanente, 1916.

Works exhibited:
"Studio", "Alla lucerna", "La casa abbandonata", "Il sofà" (?).

Brera National Exhibition, Milan, 1916.

Work exhibited:
"Il sofà" (pastel).

Gianni Maimeri Exhibition, Milan, Galleria Geri, 18 May - 2 June 1918.

Works exhibited:
"Al pianoforte", "Ritratto di mia madre", "Studio per un ritratto di mia madre" (pastel), "Il Bar", "Sotto la lampada", "Allo specchio", "In posa", "Scialle verde", "Tavolino in un caffè", "Al Tabarin", "Proposte", "Ombre alla lucerna", "Vettura di notte", "Sul divano", "Lettrice", "Un modello", "Figurina", "Teatro", "Studio d'espressione", "Viandante" (pastel), "Pezzente" (pastel), "Luci notturne (2)", "Studio di testa" (pastel), "Testina", "Bozzetto del Bar", "Note di colore (2)", "Otto studi per un quadro", "Sole d'inverno", "Aiuola fiorita", "Inverno", "Dopo la pioggia", "Ombre sul muro", "Cielo d'inverno", "Giallo e turchino", "Paesaggio invernale", "Marzo ventoso", "Giornata cupa", "Libecciata", "Costa azzurra", "Frangenti", "Grotte (2)", "Neve", "Giardino", "Rocce", "Neve e cielo", "Nevicata", "Torrente", "Autunno", "Pioggia", "Nebbia", "Sgelo", "Cumuli", "Pianta di fico", "Studi di paesaggio (2)", "Febbraio", "Ingresso del giardino", "Maestrale", "Tramonto sul mare", "Sulla spiaggia", "Fiori di campo", "Fiori e ninnoli", "Tre ricordi veneziani", "Nel cantiere".

Annual Exhibition of the Società Permanente, Milan, Palazzo della Permanente, 1920.

Work exhibited:
"Giardino".

Exhibition of artists from Veneto, Tuscany and Lombardy, Milan, Palazzo della Permanente, 1921.

Works exhibited:
"Nel bosco", "Bacino di San Marco", "Riva degli Schiavoni".

Brera National Exhibition, Milan, 1922.

Work exhibited:
"Nel giardino".

Art Exhibition of the Fascist Regional Union of the Fine Arts of Lombardy, Milan, 1929.

Work exhibited:
"Afa sulla Laguna".

Third Landscape Exhibition, Bologna, 1929.

Work exhibited:
"Fine d'inverno" (gold medal winner).

Landscape Museum, Pallanza, Palazzo Dugnani-Viani, May 1929.

Canals of Milan Exhibition, Milan, Palazzo della Permanente, May-June 1929.

Works exhibited:
"Via Senato", "Via Fatebenefratelli".

Members' Exhibition of watercolours, pastels and black and white works, Milan, Palazzo della Permanente, October-December 1930.

Works exhibited:
"Il Naviglo di Via Olocati", "Il Naviglio di Via Senato", "Pezzente" (pastel), "Lo scialle"(pastel).

First Exhibition at Galleria Amedei, Milan, Galleria Amedei, 22 February - 15 March 1930.

Works exhibited:
"Fine d'inverno", "Fiori" ("Febbraio" was to be exhibited, but was replaced by "Fiori").

Solo Exhibition, Milan, "Giornale dell'Arte" Hall, 18 March - 1 April 1930.

Works exhibited:
"Maternità", "Tappeto rosso", "Pelliccia nera", "Testa", "Signora al pianoforte", "La statuina", "Il caminetto", "Il bicchiere iridato", "Budda", "Uccellini", "Margherite", "La coppa", "Pesci persici", "Via Vallone", "Via Molino delle Armi", "Un giorno di nebbia", "Ponte sul Naviglio", "Via S. Damiano", "Ponte di Porta Romana", "Via Olocati", "Via Fatebenefratelli", "Il Lambretto", "Brina", "Il fiume", "Maretta a sera", "Bagnatura al Lido", "Ora pomeridiana", "Luna", "Mare grigio", "Verso sera", "Portovaltravaglia", "Erba di maggio", "Rigoglio estivo", "In giardino", "Vento", "Il Lambretto che spaglia", "Pomeriggio autunnale", "Mattino alla Barona", "Il Lambretto", "L'Olmo".

Gianni Maimeri Exhibition, Livorno, 1-30 June 1930.

Works exhibited:
"Corte Gabrielli (Venezia)", "Ombre sul prato (dintorni di Milano)", "Dormiente", "Nevicata", "Verso sera (Lido)", "Sul divano", "Ponte e chiusa sul Lambretto (dintorni di Milano)", "Lavandaie", "Testa", "Sulla spiaggia (Lido)", "Acero bianco", "Pelliccia nera", "Ombre sul muro", "Panni al sole", "Il gattino", "Neve", "Il Lambretto (dintorni di Milano)", "Nuvolaglia", "Il Mulino Blondel (dintroni di Milano)", "Pomeriggio autunnale (dintorni di Milano)", "Il Naviglio di Via Olocati a Milano", "Riva degli Schiavoni - Venezia", "Vecchio al caffè", "Vasi di fiori", "Figurina sulla spiaggia (San Remo)", "In giardino", "Riflessi", "Il Naviglio Grande", "Mestolone", "Giornata d'inverno (dintorni di Milan)", "Brina", "Fine d'inverno", "La cascatella", "In lavoro", "Fiori di campo", "Soffiatori di vetro", "Rose", "Signora al pianoforte", "Soffiatori di vetro", "Il caminetto", "Budda", "Astri della China", "Due pittori", "Erba di maggio", "La statuina", "Inverno fosco", "Eliantemi", "Beccaccini", "Il bar", "Ora pomeridiana (Lido)", "Uccellini".

Members' Exhibition of the Fine Arts and Permanent Exhibition Society, Milan, Palazzo della Permanente, April-May 1931.

Works exhibited:
"Meriggio", "Risaia".

Exhibition by painters Giovanni Lentini and Gianni Maimeri, Milan, Galleria Pesaro, April - May 1931.

Works exhibited:
"La Fonte", "Autoritratto", "Maternità", "Pelliccia nera", "Tappeto rosso", "Cassettoncino e rose di maggio", "Fiori di campo", "Pesci persici", "Uccellini", "Vetreria", "Sole d'inverno", "Il guado. Impressione", "Vento", "Al pianoforte", "Neve e bruma", "Portovaltravaglia", "Ombre azzurre", "Giornata di neve", "Fine d'inverno", "Marzo", "Impressione", "In giardino", "Lungo la spiaggia", "Verso sera", "Al Lido", "Mare grigio", "Bagnanti al Lido", "Maretta a sera. Adriatico", "Afa estiva", "Inverno", "Autunno", "Estate", "Mulino", "Mezzodì", "Rigoglio estivo", "Acque sognanti", "Pomeriggio autunnale", "Il Lambretto", "Riflessi", "Ultimi raggi", "Brina", "Neve e gelo", "Ombre sul prato" , "Nevicata", "Il caminetto", "Il bicchiere iridato", "Il Quartetto Sevcik", "La violoncellista Helisabeth Grummer", "Il pianista Josef Pembaur", "Ferruccio Busoni", "Mia madre al pianoforte. Impressione",

"Mia madre al pianoforte", "La lezione", "La statuina", "Dopo la cena", "Budda", "Astri della China", "La coppa", "Via Vallone", "Brume a sera", "Via Francesco Sforza", "Ponte dell'Ospedale", "Ponte di San Marco", "Ponte di Porta Ticinese", "Ponte di Porta Romana", "Via Senato", "Palazzo Visconti", "Via Fatebenefratelli", "Via Vallone", "Ponte di via Arena", "Ponte di via Vallone", "Giornata di nebbia", "La vecchia torre (tempo coperto)", "La vecchia torre (sole invernale)", "Via Mulino delle Armi", "Via Santa Sofia", "La Conca di via Senato", "Via Olocati", "Ponte di via Laghetto", "Ponte del Verziere", "Il letto del Naviglio Grande", "Studio per il letto del Naviglio Grande", "Il Naviglio a san Cristoforo", "Gesti e atteggiamenti di musicisti. Studi a memoria".

84th Members' Exhibition, Montecatini Terme, Palazzo delle Esposizioni, June-October 1931.

Work exhibited:
"Uccellini".

First National Art Quadrennial, Rome, 1931.

Work exhibited:
"Autoritratto".

Exhibition of painters Leonardo Bazzaro, Galileo Chini and Gianni Maimeri, Milan, Galleria Pesaro, January 1932.

Works exhibited:
"Nello studio del pittore", "Interno", "Coppa di Murano", "Anitra e crisantemi", "Lo specchio illuminato", "Ortaggi", "Scatola giapponese", "Rose e papaveri", "Vaso di Murano", "Mele cotogne", "Eliantemi", "Vaso e bicchiere", "Rose spampanate", "Ottobre", "Giaggioli e rose", "Beccaccini", "Vaso azzurro e rose", "Quaglie", "Libri e fiori", "Fiori di sambuco", "La mia pipa", "Carpa", "Vasetto di fiori", "Fiori gialli", "Idolo cinese", "Isola di San Giorgio", "Canale della Giudecca", "Luminaria", "La luna", "Via deserta", "Tavolini", "Caffè notturno", "Il lampione", "Suburbio", "Giostra", "Piazzetta San Marco", "Giardini pubblici", "Rio", "Riva degli Schiavoni (pioggia)", "L'uomo del bar", "Canal Grande", "Laguna", "Canale", "Riva degli Schiavoni (sole)", "Vicolo Pattari", "Corso Buenos Ayres", "Mezzanotte", "Lago di Lugano", "Verso mattina", "Lampioni e rotaie", "Alla Fiera", "La vettura", "Naviglio in secco", "Lungo il Naviglio", "Mercante ambulante".

3rd Art Exhibition of the Fascist Regional Union of the Fine Arts of Lombardy (Brera Biennial organised by the Permanente), Milan, February-March 1932.

Works exhibited:
"Tramonto" (replaced "Interno illuminato"), "Cascina alla Barona".

Members' Exhibition of the Permanente, Milan, Palazzo della Permanente, April-May 1932.

Works exhibited:
"Strada campestre", "I due canali", "Nevicata", "Ruscello tra la neve".

2nd Art Exhibition at Montecatini Spa promoted by the Fine Arts Society of Florence, July-October 1932.

Works exhibited:
"Paesaggio di neve", "Figura".

Self-Portrait Exhibition, Fine Arts and Permanent Exhibition Society, Milan, Palazzo della Permanente, November-December 1932.

Work exhibited:
"Autoritratto".

Exhibition of Modern Italian Painters organised by the Nuova Vita Cultural Institute, Milan, October-November 1933.

Works exhibited:
"Rose candide", "Interno illuminato", "Rivo".

Spring Members' Exhibition of the Permanente, Milan, Palazzo della Permanente, 1934.

Works exhibited:
"Rosa thea", "Carpa", "Fiori di Tagete", "Crepuscolo sul Lambro".

5th Exhibition of the Interprovincial Fascist Union of the Fine Arts of Lombardy, Milan, 1934.

Work exhibited:
"Fiori".

4th Exhibition of Landscape Art, Bologna, Exhibition Centre, May 1934.

Works exhibited:
"Inverno (Paesaggio lombardo)", "Acque sognanti", "Risaia".

First General Exhibition, 21 Contemporary Artists, Genoa, Galleria Vitelli, November 1934.

Works exhibited:
"Inverno lombardo", "Ruscello", "La bottiglia", "La statuina", "Fiori".

2nd National Quadrennial of Rome (winter session), Rome, January 1935.
Work exhibited:
"Il Po a Monticelli".

Members' Exhibition, Milan, Palazzo della Permanente, February-March 1935.

Works exhibited:
"I due canali", "Lavandaie al ruscello", "Carte e chicchere sul tavolino" (winner of the Sallustio Fornara prize).

Competition on War and Victory promoted by Her Majesty the Queen, March 1935.

Work exhibited:
"Generale Chinotto" (not entered in the competition, but won a prize).

6th Exhibition of the Interprovincial Fascist Union of the Fine Arts of Lombardy, Milan, Palazzo della Permanente, May 1935.

Work exhibited:
"Lo stagno dei pesci rossi".

Members' Exhibition. Milan, Palazzo della Permanente, December 1935.

Works exhibited:
"Po a Monticelli", "Siepe di Sambuco".

Members' Spring Exhibition, Milan, Palazzo della Permanente, April-May 1936.

Works exhibited:
"Strada rustica", "Fiori di primavera".

7th Exhibition of the Interprovincial Fascist Union of the Fine Arts of Lombardy, Milan, Palazzo della Permanente, 1936.

Work exhibited:
"Neve" (purchased for the Museum of Milan).

Italian Art Exhibition, World's Fair, Paris, 1936.

Work exhibited:
"Cassettoncino e rose di maggio".

Members' Spring Exhibition, Milan, Palazzo della Permanente, 1937.

Works exhibited:
"Astri Selvatici", "La chiesa della Barona".

Members' Fall Exhibition, Milan, Palazzo della Permanente, 1937.

Work exhibited:
"Sotto il fogliame".

Members' Spring Exhibition, Milan, Palazzo della Permanente, March-April 1938.

Works exhibited:
"Studio del pittore", "Il piccolo pittore", "Neve nel bosco".

Members' Fall Exhibition, Milan, Palazzo della Permanente, 1938.

Works exhibited:
"Visciole", "Il giardino Melzi a Porta Romana", "Ciliegie".

Solo exhibition by the painter Gianni Maimeri, Milan, Galleria Gian Ferrari, 16-25 June 1939.

Works exhibited:
"Autoritratto" 1930, "Interno di notte", "Estate (paesaggio lombardo)", "Ore pomeridiane", "Giardino", "Neve e gelo", "Maggio lombardo", "Luci nel bosco", "Quiete", "Lambretto", "Nebbia", "Il Po al ponte della Becca", "Ticino", "Lungo il prato (paesaggio lombardo)", "Stanza di un solitario", "Coppa di Murano", "La melarancia", "Carpa e bottiglia", "Tavola di cucina", "Pesci di lago", "Quaglie", "Tinca", "Agone", "Poiana", "Beccaccini", "Gamberi", "Sardine e triglie", "Fringuelli e verdoni", "Mele e pannocchie", "Ciliegie", "Albicocche", "Astri della China", "La zinnia rossa", "La tovaglia e le rose", "Fiori di sambuco", "Fiordalisi", "Quattro rose", "Folto di rose", "Rose thee", "Tageti", "Ruscello", "Inverno", "Paesaggio", "Naviglio".

Twelve Contemporary Artists, Milan, Galleria Nova, November 1939.

Works exhibited:
"Mezzogiorno", "Cassettoncino e rose di maggio", "Paesaggio lombardo".

Gianni Maimeri, Milan, Galleria Gian Ferrari, 29 October - 8 November 1941.

Works exhibited:
"Lo stagno dei pesci rossi", "Al margine del bosco", "Il mio rame", "Frutta sulla tovaglia", "Ribes e lamponi", "Cartoccio di marroni", "Peperoni d'Abruzzo", "Il vaso di anemoni", "Il vassoio di cristallo", "Pere mature", "Pesci e bocce", "Il vaso turchino", "Peperoni di Voghera", "Pioppi di S. Zenone Po", "Il Po a Portalbera", "Castello di Arena Po", "Lago Maggiore visto dal Mottarone", "Viale di Stradella", "Il Po presso Arena", "Catena del Rosa vista dal Mottarone", "Foce dell'Olona", "Isola di S. Zenone all'alba", "Sponda del Po a Portalbera", "Foce della Versa", "Dal Ponte di Spessa", "Il Lambro a Pontesesto", "Pontesesto", "Risaia di Bazzana", "Ponte di barche sul Po", "Il Po a S. Cipriano (a valle)", "Il Po a S. Cipriano (a monte)", "Ponte di Spessa", "Isola di San Zenone", "Il Po dal ponte di Spessa", "Alba sul Po", "Il Po verso Parpanese", "Prato al Colomberotto", "Risaia a Selvanesco", "Carro di fieno al Colomberotto", "Anitre al sole", "Tramonto a Pontesesto", "Pioppo al Beldiletto", "Colline a Stradella", "Lanca di Portalbera", "Montù Beccaria", "Chiesa di Gnignano", "L'Adda dal ponte di Rivolta", "L'Adda presso Cervignano", "Il Ticino a Vigevano", "Prateria a Bisnate".

Solo Show and Canal Commemoration Show by Gianni Maimeri, Milan, Galleria Salvetti, 7 December 1948 - 2 January 1949.

Works exhibited:
"Autoritratto", "Carrozza in attesa", "Ritratto della madre al pianoforte", "Il bicchiere iridato", "Rose e giunchiglie", "Gamberi", "Uova", "Asparagi e ramolacci", "Istrumenti musicali. Violini.", "Istrumenti musicali. Ottoni", "Tavola con verdura e frutti", "Teschio", "Cotogne e verdure", "Triglie e sarde", "Sfere", "Due pesci", "Piatto d'uva", "Funghi e verdoni", "Tavola con rame e uva", "Conchiglia e bicchiere", "Mulino alla Barona", "Brinata sul Lambro", "Canale tra gli alberi", "Pioppaia", "Barca sul Po", "Terrazza sul Pincio", "Acque sognanti", "Siepe di sambuco", "L'Adda dal ponte di Rivolta", "Po in deflusso", "Mergellina", "Piazzale sul Gianicolo", "Marina a Monterosso", "Acque tra gli scogli", "Spiaggia a Monterosso", "Tramonto", "Bagnanti", "Vernazza", "Barche nel Porto di Monterosso", "Ultimo meriggio", "Mareggiata", "Tetto rosso".

Canals in Milan (thirty views from dawn to sunset, in February and March 1929): "Vecchio ponte di via Vallone", "Ponte del Verziere", "Il ponte di via Arena e la Conca di Leonardo", "Naviglio di S. Sofia", "Lungo la via Francesco Sforza", "Il ponte delle Sirenette", "Ponte Marcellino e Conca di S. Marco", "Il ponte e le vecchie case di via Vallone", "Meriggio sul Naviglio di via Senato", "Palazzo Visconti a S. Damiano", "Giornata grigia al Mulino delle Armi", "Ponte di S. Celso", "Via Olocati e la Torre del Sale", "La Conca di via Senato", "Sostra di legna sul Naviglio di S. Sofia", "Ponte dell'Ospedale", "Il ponticello del Laghetto dell'Ospedale", "Giorno di nebbia (Naviglio del Vallone)", "Via Vallone", "Alba sul Naviglio di via Senato", "Il ponte del Verziere", "Vecchie case sul ponte del Verziere", "Mulino", "Ponte di Porta Romana", "Naviglio lungo la via Fatebenefratelli", "Ponte di Porta Ticinese", "Tramonto a S. Sofia", "L'ultima barca sul Naviglio", "La torre del Sale", "Brume a sera", Sketches from Life: Musicians: From Busoni to Toscanini (1905-1946).

Solo show by Gianni Maimeri, Milan, Galleria Gian Ferrari, June 1949.

Work exhibited:
"Coppa di Murano".

Exhibition by Gianni Maimeri, Rome, Galleria La Barcaccia, 3-14 November 1949.

Works exhibited:
"Capo Milano in Ischia", "Il Porto d'Ischia", "Il Porto d'Ischia dal Belvedere", "Spiaggia di pescatori", "Il Castello di Procida", "Barche al Porto d'Ischia", "Sulla scalea della Pagoda", "Dalla terrazza di casa Bavarese", "Forio d'Ischia", "Dal Belvedere", "Spiaggia a Porto d'Ischia", "Albergo Ischia", "Via Giacinto Gigante", "Salvatore padre", "Processione a Ischia", "Barche al molo di Porto d'Ischia", "Panorama verso Procida", "Il Castello di Procida", "Anna dal Molo di Porto d'Ischia", "Il Castello d'Ischia", "La polverosa strada di Barano", "Molo principale di Porto d'Ischia", "Orme di cavallo sulla strada", "Spiaggia d'Ischia", "Chiesa di San Pietro", "Temporale a Ischia", "Molo di Procida al tramonto", "La spiaggia dei Maroniti", "Autoritratto", "Carrozza in attesa", "Il bicchiere iridato", "Al caminetto", "Interno", "Uova", "Sfere", "Due pesci", "Piatto d'uva", "Tavola con frutti e rame", "Tavola con uva e rame", "Funghi e verdoni", "Teschio", "Canale tra gli alberi", "Mareggiata a Monterosso", "Casina settecentesca sul Naviglio", "Risaia in primavera", "Monterosso al mare", "Naviglio grande a Robecco", "Lambro meridionale", "Lanca sul Po", "L'Adda dal ponte di Rivolta", "Po in deflusso", "La Fontana di Villa Medici".

Solo show by Gianni Maimeri of Landscapes on the Island of Ischia and Impressions of Rome, Galleria d'Arte Travaglini, Milan, 1-15 February 1950.

Works exhibited:
"S. Angelo", "Il Porto di Ischia", "Il Porto di Ischia all'imbrunire", "Case sulla Riva dei Pescatori", "Il lido di Porto d'Ischia", "Chiesa di Ischia - Ponte", "Viale fiorito", "La carrozzella", "Baia di S. Montano", "Sulla spiaggia", "Case sul molo di Procida", "Molo di Procida", "Chiesa di S. Pietro a Ischia", "Bacino del porto a sera", "Marina verso Procida", "Pineta", "Barche all'ormeggio", "Sulla strada di Barano", "Forio d'Ischia", "Il Porto a sera", "La foce dell'Olmitello", "Velieri all'ormeggio", "Velieri allo scarico", "Barconi nel porto", "Piazza S. Silvestro", "Piazza del Campidoglio", "Ponte Cestio", "Ponte Nomentano sull'Aniene", "Colonne del Tempio di Venere Genitrice", "L'Aniene al Monte Sacro".

6th "Amaro Ramazzotti" National Painting Competition, Milan, Palazzo Sociale, 13-30 June 1959.

Works exhibited:
"I Navigli di Milano", "Vecchio ponte di via Vallone", "Via Olocati e la Torre del Sale", "Ponte del Verziere", "La Conca di via Senato", "Il ponte di via Arena e la Conca di Leonardo", ""Sciostree" sul Naviglio di S. Sofia", "Naviglio di S. Sofia", "Ponte dell'Ospedale", "Lungo la via Francesco Sforza", "Il ponticello del Laghetto

dell'Ospedale", "Il ponte delle Sirenette", "Giorno di nebbia sul Vallone", "Ponte Marcellino a S. Marco", "Via Vallone e Albergo Popolare", "Il ponte e le vecchie case di via Vallone", "Alba sul Naviglio di via Senato", "Meriggio sul Naviglio di via Senato", "Il ponte del Verziere", "Palazzo Visconti a S. Damiano", "Vecchie case sul ponte del Verziere", "Giornate grigie al Mulino delle Armi", "Il Naviglio al Mulino delle Armi", "Ponte di S. Celso", "Ponte di Porta Romana", "Naviglio lungo la via Fatebenefratelli", "L'ultima barca sul Naviglio di via Senato", "Ponte di Porta Ticinese", "La Torre del Sale", "Tramonto a Santa Sofia", "Brume a sera".

Commemorative Exhibition of Member Painters, Milan, Palazzo della Permanente, 9 April - 10 May 1960.

Works exhibited:
"Mia madre", "Le due amiche", "Il Lambretto d'autunno", "Cassettoncino e rose di maggio", "Bocce e uova", "Il mio studio", "Il bruum", "Autoritratto", "Bragozzi a Venezia", "Siepe alla Barona", "Pesci".

First Exhibition of Late Artists, Milan, Palazzo del Turismo, 12-28 October 1963.

Works exhibited:
"Monterosso al mare", "Paesaggio lombardo", "Venezia".

Third Exhibition of Late Artists, Piacenza, Great Hall in Palazzo Gotico, 4-19 April 1964.

Work exhibited:
"L'ansa del Po".

Fourth Exhibition of Late Artists, Verona, Palazzo della Gran Guardia, 14-31 May 1964.

Sixth Exhibition of Late Artists, Florence, Fine Arts Society - "Casa di Dante" Artists' Circle, 2-15 January 1965.

Works exhibited:
"Bacino di San Marco (Venezia)", "Beccaccini", "Rose".

Seventh Exhibition of Late Artists, Pavia, Great Hall in Castello Visconteo, 15-30 April 1965.

Works exhibited:
"Rose bianche", "Bacino di San Marco a Venezia", "Ruscello alla Barona", "Cesto di frutta", "Ponte di Rialto a Venezia", "Disegni della raccolta", "Celebri musicisti".

Exhibition of Late Artists, Treviglio, Upper Hall in the Cassa Rurale ed Artigiana, 15-27 May 1965.

Ninth Exhibition of Late Artists, Milan, Palazzo del Turismo, 5-25 February 1966.

Works exhibited:
"Strumenti musicali", "Primavera alla Barona", "Venezia", "Le rose", "Tramonto a Porto d'Ischia".

Exhibition of Late Artists, Clusone, Elementary Schools, August 1966.

Work exhibited:
"Lanca del Po".

Art Exhibition, Milan, Palazzo del Turismo, March 1967.

Works exhibited:
"Bocce, biglie e uova", "Paesaggio", "Natura morta".

Art Exhibition, Milan, Palazzo del Turismo, 3-18 February 1968.

8th Sketch Exhibition, Vicenza, Galleria del Bacchiglione, 22 February - 5 March 1969.

Works exhibited:
"Il Bar", "Porto d'Ischia", "Scoglio a Monterosso al mare", "Barche a Ischia".

Exhibition of Italian and Italo-Argentinian Painters, Rome, Palazzo Firenze, 10-20 December 1969.

Works exhibited:
"Venezia", "Natura morta - pesci".

27th Art Exhibition, Milan, Visconti Castle, 3-20 May 1971.

Works exhibited:
"Cortile lombardo", "Lanca del Po", "Il bosco", "Il mio gattino", "Il Po a San Zenone", "Ciliegie", "Monterosso al mare", "Osteria (Ca Bianca)", "Pesci".

Musicians as seen by Gianni Maimeri, Milan, Museum of Teatro alla Scala, 1-31 October 1992.

Musicians as seen by Gianni Maimeri, Ancona, Palazzo degli Anziani, 5 November - 6 December 1992.

Musicians and Flowers as seen by Gianni Maimeri, Spello, Vecchio Palazzo Comunale, 29 May - 19 June 1993.

Works exhibited:
"Donna distesa con bouquet di fiori", "Fiori di zucca", "Consolle e vaso con rose bianche e rosse", "Vasi di fiori", "Vaso con fiori", "Bozzetto per vaso di fiori", "Vaso con mimose", "Vaso con fiori di campo", "Anemoni rossi, rosa, gialli e cacciagione", "Cineserie e vaso con fiori", "Vaso di anemoni", "Vaso con tre rose gialle", "Vaso con mimosa", "Mazzo di garofani", "Vaso con fiori", "Vassoio con ramo di glicine", "Bozzetto per vaso di rose", "Vaso cinese con fiori", "Fiori", "Vaso con rose gialle e rosa", "Vaso con ortensie azzurre e rosa", "Vaso con gigli bianchi, iris e conchiglia", "Statua con rose", "Vaso con garofani rossi", "Vaso con primule", "Vaso con ortensie bianche", "Vaso con genziane", "Vaso con fiori e statua", "Vaso con rose gialle", "Vaso di fiori con uccelli e farfalle", "Vaso con fiordalisi, rose, fiori di campo e farfalle", "Fantasia notturna di fiori", "Vaso con rose gialle", "Gigli bianchi", "Orchidee", "Tavola con fiori, frutta e verdura", "Vaso con narcisi, anemoni e violette", "Ciclamini e ortaglia", "Ciclamini e funghi", "Vaso con rose gialle e garofani rossi", "Vaso con garofani, rose e mimose", "Vaso con fiori di campo", "Bosco con arbusti di rose rosa", "Vaso di terracotta con fiori", "Vaso di gladioli", "Bozzetto per vaso di fiori", "Vaso con ortensie azzurre e rosa", "Vaso con due rose", "Vaso con tre rose", "Vaso con due rose", "Vaso con fiori di campo".

Musicians as seen by Gianni Maimeri, Nocera Umbra (Perugia), Sala dell'Esposizione, 23 July - 21 August 1994.

Landscapes, Still Lifes and Portraits, Nocera Umbra (Perugia), Sala Consiliare del Palazzo Comunale, 23 July - 21 August 1994.

Works exhibited:
"Moglie e figli alla Carlesca", "Il guanto", "Paesaggio lombardo al tramonto", "Tabarin", "Schizzo per il Tabarin", "Lambretto alla Barona", "Gli archi", "Gli ottoni", "Lo scialle", "Bozzetto per il Tabarin", "Mia moglie nel parco a Portovaltravaglia", "La mia scrivania", "Vaso con fiori", "Attraversando la Carlesca", "Mare d'inverno", "Bocce, biglie e uova", "Autoritratto", "Il guanto", "Le amiche", "Partita a poker", "Nevicata sul Lambretto alla Barona", "Paesaggio", "Bacino di San Marco a Venezia", "Il mio studio alla Barona", "Lido di Venezia", "Venezia. Riva degli Schiavoni", "Mezzanotte", "Mio figlio al cavalletto", "Consolle con vaso di rose".

Landscapes, Still Lifes and Portraits, Montefalco (Perugia), Enoteca Coccorone, 1-20 August 1994.

Works exhibited:
"Moglie e figli alla Carlesca", "Il guanto", "Paesaggio lombardo al tramonto", "Tabarin", "Schizzo per il Tabarin", "Lambretto alla Barona", "Gli archi", "Gli ottoni", "Lo scialle", "Bozzetto per il Tabarin", "Mia moglie nel parco a Portovaltravaglia", "La mia scrivania", "Vaso con fiori", "At-

traversando la Carlesca", "Mare d'inverno", "Bocce, biglie e uova", "Autoritratto", "Il guanto", "Le amiche", "Partita a poker", "Nevicata sul Lambretto alla Barona", "Paesaggio", "Bacino di San Marco a Venezia", "Il mio studio alla Barona", "Lido di Venezia", "Venezia. Riva degli Schiavoni", "Mezzanotte", "Mio figlio al cavalletto", "Consolle con vaso di rose".

Musicians as seen by Gianni Maimeri and Oils. Landscapes, Still Lifes and Portraits, Strasbourg, European Parliament, 10-13 October 1995.

Works exhibited:
Drawings from the series "I Musicisti", "Bozzetto per il Tabarin", "Bozzetto per il Tabarin", "Le amiche", "Bacino di San Marco a Venezia", "Laguna da Piazza di San Marco", "Ritratto di una signora (Anna)", "La lavandaia", "Nevicata sul Lambretto alla Barona", "Carro da fieno", "Il fosso della lavandaia", "Autoritratto", "Il mio studio alla Barona", "Mio figlio al cavalletto", "Bocce, biglie e uova", "Il Castello d'Ischia al tramonto", "Cupola della chiesa a Porto d'Ischia", "La mia scrivania", "Gladioli", "Vaso di rose thea", "Paesaggio", "Moglie e figli alla Carlesca", "Il guanto", "Paesaggio lombardo al tramonto", "L'estate", "Tabarin", "Lambretto alla Barona", "Gli archi", "Gli ottoni", "Lo scialle", "Vaso con fiori e statua", "Vaso con fiori di campo e conchiglia", "Bosco con arbusti di rose", "Attraversando la Carlesca", "Mare d'inverno", "Edwin Fischer. Concerto per pianoforte e orchestra", "Autoritratto", "Partita a poker", "Paesaggio", "La mia scrivania", "Lido di Venezia", "Venezia. Riva degli Schiavoni", "Il salice piangente".

Musicians as seen by Gianni Maimeri and Oils. Landscapes, Still Lifes and Portraits, Reggio Emilia, 13th Antiques Fair, Exhibition Centre, 26 January - 4 February 1996.

Works exhibited:
Drawings from the series "I Musicisti", "Bozzetto per il Tabarin", "Bozzetto per il Tabarin", "Le amiche", "Bacino di San Marco a Venezia", "Laguna da Piazza di San Marco", "Ritratto di una signora (Anna)", "La lavandaia", "Nevicata sul Lambretto alla Barona", "Carro da fieno", "Il fosso della lavandaia", "Autoritratto", "Il mio studio alla Barona", "Mio figlio al cavalletto", "Bocce, biglie e uova", "Il Castello d'Ischia al tramonto", "Cupola della chiesa a Porto d'Ischia", "La mia scrivania", "Gladioli", "Vaso di rose thea", "Paesaggio", "Moglie e figli alla Carlesca", "Il guanto", "Paesaggio lombardo al tramonto", "L'estate", "Tabarin", "Lambretto alla Barona", "Gli archi", "Gli ottoni", "Lo scialle", "Vaso con fiori e statua", "Vaso con fiori di campo e conchiglia", "Bosco con arbusti di rose", "Attraversando la Carlesca", "Mare d'inverno", "Edwin Fischer. Concerto per pianoforte e orchestra", "Autoritratto", "Partita a poker", "Paesaggio", "La mia scrivania", "Lido di Venezia", "Venezia. Riva degli Schiavoni", "Il salice piangente".

Presentation of Gianni Maimeri's Diaries, Milan, Fondazione Maimeri, 18 January 1997.

Flowers. Un fiore per la vita, Rome, 25 January 1997.

The Erotic Imagination: Sensuality and voyeurism in 19th and 20th century painting, Reggio Emilia, Antiques Fair, 24 January - 2 February 1997.

Works exhibited:
"Nudo sul divano verde", "Studio di nudo sul divano".

Selected works, Varese, Villa Recalcati, 10 April - 23 May 1997.

Works exhibited:
"Nudo di schiena", "Tabarin", "Bozzetto per il Tabarin", "Figura nuda in movimento", "Lo scialle", "Mia moglie", "Fiori di zucca", "Leda e il cigno", "Nudo sul divano", "L'Orchestra", "Venezia. Riva degli Schiavoni", "Mia moglie nel parco a Portovaltravaglia", "Mia moglie in blu", "Le amiche", "Bacino di San Marco a Venezia",

"Il guanto", "Anna", "Donna con scialle giallo", "Mia moglie in giardino con ortensie", "Donna con bimbo", "La lavandaia", "Nevicata sul Lambretto alla Barona", "Spiaggia al Lido di Venezia", "Carro da fieno", "I Navigli di Milano", "L'estate", "Autoritratto", "Attraversando la Carlesca", "Fosso alla Barona", "Mio figlio al cavalletto", "Igor Stravinsky", "Branislaw Hubermann", "Pietro Mascagni", "Carlo Vidusso", "Antonio Janigro", "Lilia d'Albore", "Vaso con due rose", "Nicolaj Orloff", "Luigi Chiarappa", "Gregor Piatigorski", "Pina Carminelli", "Mia figlia al pianoforte", "Strumenti ad arco", "Gli ottoni", "Arturo Toscanini", "Il porto di Ischia dalla strada del Belvedere", "Pineta al porto di Ischia", "Gilberto Crepax", "Donna in azzurro", "Paesaggio lombardo al tramonto", "Lambretto alla Barona".

Selected works, Asolo, Civic Museum, 28 August - 10 October 1997.

Works exhibited:
"Nudo di schiena", "Tabarin", "Bozzetto per il Tabarin", "Figura nuda in movimento", "Lo scialle", "Mia moglie", "Fiori di zucca", "Leda e il cigno", "Nudo sul divano", "L'Orchestra", "Venezia. Riva degli Schiavoni", "Mia moglie nel parco a Portovaltravaglia", "Mia moglie in blu", "Le amiche", "Bacino di San Marco a Venezia", "Il guanto", "Anna", "Donna con scialle giallo", "Mia moglie in giardino con ortensie", "Donna con bimbo", "La lavandaia", "Nevicata sul Lambretto alla Barona", "Spiaggia al Lido di Venezia", "Carro da fieno", "I Navigli di Milano", "L'estate", "Autoritratto", "Attraversando la Carlesca", "Fosso alla Barona", "Mio figlio al cavalletto", "Igor Stravinsky", "Branislaw Hubermann", "Pietro Mascagni", "Carlo Vidusso", "Antonio Janigro", "Lilia d'Albore", "Vaso con due rose", "Nicolaj Orloff", "Luigi Chiarappa", "Gregor Piatigorski", "Pina Carminelli", "Mia figlia al pianoforte", "Strumenti ad arco", "Gli ottoni", "Arturo Toscanini", "Il porto di Ischia dalla strada del Belvedere", "Pineta al porto di Ischia", "Gilberto Crepax", "Donna in azzurro", "Paesaggio lombardo al tramonto", "Lambretto alla Barona", "Notturno al Duomo".

From night to light, Lecce, "S. Castromediano" Museum, 22 November - 21 December 1997.

Works exhibited:
"Il guanto", "Moglie e figli alla Carlesca", "Paesaggio lombardo al tramonto", "L'estate", "Tabarin", "Anna" (pastel), "Lambretto alla Barona", "Mia figlia al pianoforte", "Gli archi", "Gli ottoni", "L'orchestra impazzita", "Vaso con due rose", "Lo scialle", "Bozzetto per il Tabarin", "Mia moglie nel parco a Portovaltravaglia", "Donna che danza (bozzetto per il Tabarin)", "Attraversando la Carlesca", "Mare d'inverno", "Fosso alla Barona", "Donna in azzurro", "Donna con scialle giallo", "Autoritratto", "Le amiche", "Nevicata sul Lambretto alla Barona", "Bacino di San Marco a Venezia", "Mia moglie" (pastel), "Lido di Venezia", "La raccolta delle mele", "Venezia. Riva degli Schiavoni", "Mezzanotte", "Mio figlio al cavalletto", "Nudo accovacciato sul divano".

Selected works, Castelfranco Veneto, Casa del Giorgione, 13 June - 20 September 1998.

Works exhibited:
"Nudo di schiena", "Tabarin", "Bozzetto per il Tabarin", "Figura nuda in movimento", "Lo scialle", "Mia moglie", "Fiori di zucca", "Leda e il cigno", "Nudo sul divano", "L'Orchestra", "Venezia. Riva degli Schiavoni", "Mia moglie nel parco a Portovaltravaglia", "Mia moglie in blu", "Le amiche", "Bacino di San Marco a Venezia", "Il guanto", "Anna", "Donna con scialle giallo", "Mia moglie in giardino con ortensie", "Donna con bimbo", "La lavandaia", "Nevicata sul Lambretto alla Barona", "Spiaggia al Lido di Venezia", "Carro da fieno", "I Navigli di Milano", "L'estate", "Autoritratto", "Attraversando la Carlesca", "Fosso alla Barona", "Mio figlio al cavalletto", "Igor Stravinsky", "Branislaw Hubermann", "Pietro Mascagni", "Carlo Vidusso", "Antonio Janigro", "Lilia d'Albore", "Vaso con due rose", "Nicolaj Orloff", "Luigi Chiarappa", "Gregor

Piatigorski", "Pina Carminelli", "Mia figlia al pianoforte", "Strumenti ad arco", "Gli ottoni", "Arturo Toscanini", "Il porto di Ischia dalla strada del Belvedere", "Pineta al porto di Ischia", "Gilberto Crepax", "Donna in azzurro", "Donna in azzurro", "Lambretto alla Barona".

Cadegliano, Viconago, 18 July - 3 August 1998.

Works exhibited:
"Lanca del Po con Anna", "Il Naviglio a Cassinetta di Lugagnano", "Mia moglie con mia figlia".

Portraits and figures, Campione d'Italia, Galleria Civica, 5 June - 25 July 1999.

Works exhibited:
"Nudo", "Tabarin", "Mezzanotte", "Lo scialle", "Il mendicante", "Mia moglie " (pastel), "Leda e il cigno", "Nudo di schiena", "Nudo sul divano verde", "La raccolta delle mele", "Nudo con fiori in grembo", "Le amiche", "Anna" (pastel), "Donne sul divano", "Donna con scialle giallo", "Il guanto", "Mia moglie con mia figlia", "Dipinto per un manifesto pubblicitario", "La lavandaia", "Ritratto di Zoe Maimeri", "Vele nella laguna", "Carro da fieno", "Autoritratto", "Ritratto di Leone Maimeri", "Mio figlio al cavalletto", "Mia moglie nel parco a Portovaltravaglia", "Giardino Melzi d'Eril", "Ritratto di Alfonso Catalano", "Frutta, verdura, rami e terracotta", "Partita di poker", "Attraversando la Carlesca", "Moglie e figli alla Carlesca".

20th century painting in Lombardy, Vigevano, Castello Sforzesco, 31 March - 1 July 2001.

Work exhibited:
"Paesaggio lombardo al tramonto".

Musicians as seen by Gianni Maimeri, Milan, Teatro dal Verme, 12 February - 1 June 2003.

From the past to the future. Gianni Maimeri (1884-1951), Milan, Galleria Le Pleiadi, 7 May - 12 June 2004.

Works exhibited:
"Le amiche", "Dormiente sul divano verde", "Mia moglie con lo scialle", "Donna seduta", "Contadina", "Baveno. Lago Maggiore", "Pianta, lago e collina", "Baveno", "Fiume", "Il Ticino a Vigevano", "Notturno a Zurigo", "Lambretto alla Barona", "La raccolta delle mele", "Il salice piangente", "Fosso verdissimo", "Primavera alla Barona", "La lavandaia", "Il giardino", "Contadino", "Paesaggio", "Strada all'isola di Ischia", "Siepe alla Barona", "Pineta a Ischia", "Spiaggia di pescatore al porto di Ischia", "Processione di barche al tramonto (Ischia)", "Barche al sole del mattino al porto di Ischia", "Archi di ponte", "Porticato al Robecco sul Naviglio", "Rio a Venezia", "Notturno al Duomo", "Vallata", "Vaso con garofani, rose e mimose", "Tovaglia di fiori con coppa di fiori", "Bozzetto per vaso di rose", "Bozzetto di vaso di fiori", "Vaso con ortensie", "Crostacei sul tomo", "Bottiglie e natura morta", "Cipolle e cavolfiore", "Natura morta (mele)", "Anitra e crisantemi", "Schizzo di vetro".

From night to light, Gravellona Lomellina, Teatro Sociale, 2-17 June 2007.

Works exhibited:
"Mia figlia al pianoforte", "Bozzetto per il Tabarin", "Strada alla Barona", "Chiesa controluce", "Bozzetto per i burattini", "Il Po dal ponte di Spessa", "Mia moglie al pianoforte", "Sabbione al Po", "Ritratto del signor Moro", "Figura seduta".

Water and light, Milan, Fondazione Maimeri, 5 December 2007.

Works exhibited:
Series of "I Navigli di Milano", "Tabarin", "Schizzo per il Tabarin", "Mulino della Barona", "Il brumista al Tabarin (Carrozza di notte)", "Lo scialle", "Le lavandaie", "Edwin Fischer. Concerto per pianoforte e orchestra", "Autoritratto", "Ne-

vicata sul Lambretto alla Barona", "Bacino di San Marco a Venezia", "Concerto per pianoforte", "Venezia. Riva degli Schiavoni", "Mezzanotte", "Notturno al Duomo".

Gola and Maimeri. Colours and sentiments of early 20th century painting in Lombardy, Modena, Modena Antiquaria, 15-24 February 2008.

Works exhibited:
"Lo studio di Emilio Gola", "Mezzanotte", "Tabarin", "Lo scialle", "Anna di profilo", "Le amiche", "Donna con scialle giallo", "Il guanto", "Le lavandaie", "Nevicata sul Lambretto alla Barona", "Mia moglie nel parco a Portovaltravaglia", "Cascina alla Barona (Cascina Ranza)", "Fosso alla Barona", "Moglie e figli alla Carlesca", "Attraversando la Carlesca", "Paesaggio lombardo al tramonto", "Spiaggia al Lido di Venezia", "Venezia. Riva degli Schiavoni", "Bacino di San Marco", "Mare d'inverno".

Gola and Maimeri. Colours and sentiments of early 20th century painting in Lombardy, Milan, Fondazione Maimeri, 29 February - 16 March 2008.

Works exhibited:
"Lo studio di Emilio Gola", "Mezzanotte", "Tabarin", "Lo scialle", "Anna di profilo", "Le amiche", "Donna con scialle giallo", "Il guanto", "Le lavandaie", "Nevicata sul Lambretto alla Barona", "Mia moglie nel parco a Portovaltravaglia", "Cascina alla Barona (Cascina Ranza)", "Fosso alla Barona", "Moglie e figli alla Carlesca", "Attraversando la Carlesca", "Paesaggio lombardo al tramonto", "Spiaggia al Lido di Venezia", "Venezia. Riva degli Schiavoni", "Bacino di San Marco", "Mare d'inverno".

Maimeri and the Courtyards, Milan, Fondazione Maimeri, May-June 2008.

Works exhibited:
"Porticato a Robecco sul Naviglio", "Porticato a Robecco", "Il Giardino", "Le Lavandaie", "Nevicata sul Lambretto alla Barona", "Fosso alla Barona", "Chiesa alla Barona", "Moglie e figli alla Carlesca", "Paesaggio", "Giardino Melzi d'Eril", "La siepe", "Finestra alla Barona".

20th century masters. Ideas for a picture gallery, Legnano, Castello di San Giorgio, 3 May - 6 July 2008.

Work exhibited:
"Attraversando la Carlesca".

Gianni Maimeri: a critical bibliography

MAITE ROSSI AND PAOLA TRAVAGLIO

Below is a list, in chronological order, of critical publications and catalogues of exhibitions of the works of Gianni Maimeri.

Esposizione di Gianni Maimeri (Milan, Galleria Geri), R. GIOLLI (ed.), Milan 1918.

Prima Mostra della Galleria Amedei, Milan 1930.

"Mostra Gianni Maimeri" (Livorno, "Bottega d'Arte" Gallery), F. VIVIANI (ed.), in *Bollettino di Bottega d'Arte*, Livorno 1930, IX, 8.

Mostra dei pittori Giovanni Lentini e Gianni Maimeri (Milan, Galleria Pesaro), G. BOTTA (ed.), Milan-Rome 1931.

LXXXIV Esposizione Sociale (Montecatini Terme, Palazzo delle Esposizioni), Montecatini Terme 1931.

Mostra dei pittori Leonardo Bazzaro, Galileo Chini e Gianni Maimeri (Milan, Galleria Pesaro), Milan-Rome 1932.

"IV Esposizione dell'Arte del Paesaggio" (Bologna, Fiera Esposizione), in *Il volto della patria,* magazine published by the Associazione Nazionale per Paesaggi e Monumenti Pittoreschi d'Italia, Bologna 1934.

Prima Mostra Complessiva. 21 Artisti Contemporanei (Genoa, Galleria Vitelli), Milan 1934.

Mostra Personale del Pittore Gianni Maimeri (Milan, Galleria Gian Ferrari), G. BOTTA (ed.), Milan 1939.

Dodici Artisti d'Oggi (Milan, Galleria Nova), Milan 1939.

G. BOTTA, *Gianni Maimeri* (Milan, Galleria Gian Ferrari), Milan 1941.

G. BOTTA, *Gianni Maimeri. Tre saggi critici*, Milan 1948.

Mostra Personale e Mostra Commemorativa del Naviglio di Gianni Maimeri (Milan, Galleria Salvetti), G. PREDAVAL (ed.), Milan 1949.

Mostra del pittore Gianni Maimeri (Rome, Galleria "La Barcaccia"), M. BIANCALE (ed.), Rome 1949.

6th "Amaro Ramazzotti" National Painting Competition (Milan, Palazzo Sociale), G. RAMAZZOTTI (ed.), Milan 1959.

Mostra Commemorativa di Pittori Soci (Milan, Palazzo della Permanente), R. TACCANI (ed.), Milan 1960.

Prima Mostra Artisti Scomparsi (Milan, Palazzo del Turismo), presented by UIVA, Milan 1963.

Terza Mostra Artisti Scomparsi (Piacenza, Salone del Palazzo Gotico), presented by UIVA, Milan 1964.

Sesta Mostra Artisti Scomparsi nell'ultimo cinquantennio (Florence, Società delle Belle Arti, Circolo degli Artisti "Casa di Dante"), presented by UIVA, Florence 1965.

Settima Mostra Artisti Scomparsi (Pavia, Salone del Castello Visconteo), presented by UIVA, Pavia 1965.

Mostra d'Arte (Milan, Palazzo del Turismo), presented by UIVA, Milan 1967.

Mostra di Pittori Italiani e Italo Argentini (Rome, Palazzo Firenze), presented by UIVA, Rome 1969.

XXVII Mostra d'Arte (Pavia, Castello Visconteo), presented by UIVA, Milan 1971.

A. MANDELLI, *Toscanini: Appunti per un bilancio critico*, Milan 1972.

Gianni Maimeri (1884-1951). Dal notturno alla luce, G. BUCCELLATI (ed.), Milan 1991.

E. BORRUSO, *Arte e Industria*, Milan 1991.

Un Maestro del colore nella Pittura del Novecento italiano. I fiori di Gianni Maimeri (1884-1951) (Spello, Vecchio Palazzo Comunale), R. DE GRADA, P. MANAZZA AND S. ZUFFI (eds.), Milan 1993.

Arte è... Collettiva di Pittori Contemporanei (Montefalco, Palazzo dei Priori), Montefalco 1994.

Un Maestro del colore nella Pittura del Novecento italiano. Gianni Maimeri (Strasbourg, European Parliament), Vimodrone 1995.

I diari di Gianni Maimeri, G. BUCCELLATI (ed.), Milan 1996.

Gianni Maimeri (1884-1951). Opere scelte, R. DE GRADA AND P. MANAZZA (eds.), Milan 1997.

L'Immaginario Erotico. Sensualità e voyeurismo nella Pittura tra Ottocento e Novecento, A. FIZ, P. MANAZZA, P. NICHOLLS (eds.), Milan 1997.

Gianni Maimeri (1884-1951). Ritratti e figure (Campione d'Italia, Galleria Civica), Milan 1999.

La pittura in Lombardia nel XX secolo (Vigevano, Castello Sforzesco), R. DE GRADA (ed.), Cinisello Balsamo 2001.

I Musicisti visti da Gianni Maimeri ritratti tra il 1911 e il 1946 (Milan, Teatro dal Verme), Milan 2003.

Dal passato al futuro. Gianni Maimeri (1884-1951) (Milan, Galleria d'Arte Le Pleiadi), R. DE GRADA (ed.), Bettolino di Mediglia 2004.

Acqua e luce. Milan tra Ottocento e Novecento. Navigli e Notturni, G. BUCCELLATI AND B. MANETTI (eds.), Milan 2007.

Gola e Maimeri. Colori e sentimenti protagonisti della pittura lombarda nel primo '900, R. DE GRADA, L. LUALDI AND P. MANAZZA (eds.), Milan 2008.

Ottocento. Catalogo dell'arte italiana dell'Ottocento, L. LUALDI (ed.), 37, Milan 2008.

Italian publications and manuals on painting techniques (1900-1940)

PAOLA TRAVAGLIO

This brief contribution following the publication of Gianni Maimeri's *Treatise on Paint* is intended to clarify the bibliographic context of works on artistic techniques available in Italy at the time he wrote his treatise.

I think that if we are to make progress in research into the sources available for study of artistic techniques and better understand even a single work such as the treatise published herein, it may be of use first of all to provide a bibliographic overview, necessarily incomplete in our case. I have attempted to divide up the chronological sequence of publications by theme, to promote better understanding of the complexity and variety of treatises, above all in consideration of and for the use of various specialists converging on this field (art historians, restoration workers, chemists, artists and painters, students in schools of the fine arts).

Silvia Bordini's excellent volume[1] surveying the vast historical bibliography of artistic techniques ends its bibliographic review at the end of the nineteenth century, by the author's own explicit choice, without covering any of the issues and debates of the twentieth century. But it is in the twentieth century that we find the contrast between tradition and innovation, a focus on the true meaning of artistic techniques (especially painting techniques), the advent of new materials and new techniques (such as those involving use of synthetic resins), and diversification of ideological and stylistic indicators which are quite different from one another, destined to break up the transmission of knowledge that historically always presented a practically unitary character. This makes it even more important to look into the theoretical debate surrounding technical artistic know-how in this century.

Despite the numerous studies of artistic technique that have been published in recent years, this field of study still remains entirely to be discovered. In fact, reading collections of artistic literature, reports on restoration, historical and artistic treatises and art dictionaries and encyclopedias, we get the impression that the study of technique, especially in the twentieth century and particularly in Italy, is given very little consideration, or that no treatises or formularies were published in this period. In actual fact, as Francesca Tolaini writes, "This is the story of a wealth of volumes that were published and then forgotten, of editions written and rewritten, of discoveries that went unobserved. It is true that many steps remain yet to be taken, before these studies may obtain more complete, organic results, but it is also true that the greater part of the work continues to be unknown to the majority."[2] For an overview of the technical and artistic precepts of the western nations, the reader is referred to the precious bibliography of about one hundred and fifty titles appearing at the end of the work by Maria Bazzi, *Abecedario pittorico*,[3] covering the first half of the twentieth century. Much information may also be found in works of general interest, which we shall not describe here, such as formularies and encyclopedias, for instance Italo Ghersi's two volumes, *Ricettario Domestico* and *Ricettario Industriale*,[4] or *1300 ricette utili per la casa* by Luigi Solari.[5]

A work which is still very useful on a European scale, though it often lacks information on the date and source of the information, is the *Dictionnaire des secrets d'artisans disparus*, a work published in four volumes in Brussels starting in 1995. On pigments, in addition to the works we will discuss below, it is always useful to consult chemical texts such as Vignola's *Analisi dei pigmenti minerali* (part of a series of "Technical-chemical monographs of varnishes and colorants" published by Umberto Cuppini, Milan 1947) and the *Dizionario di merceologia e chimica applicata* by Villavecchia-Eigenmann,[6] inspired by works and essays of a strictly chemical and analytic nature published by the great Italian chemist, which dedicates a whole section to the analysis and description of the principal pigments and colorants used in the first half of the century.

Painting Technique

Of great interest on the topic of painting techniques is the conspicuous production of Raffaele Armando Califano-Mundo, the Neapolitan author of *Pittura ad alluminuria e pastello, Manuale della pittura a guazzo e a tempera* (Naples 1910), *Manuale della pittura encaustica e a fresco, Manuale della pittura a smalto, Manuale della pittura su vetro, Manuale della pittura ad olio* (Naples 1910), *Dizionarietto degli utensili e delle sostanze adoperate in pittura, Manuale di pittura e acquarello - miniatura e restauro* (Naples 1910).

This series of small popular handbooks of good quality may be said to cover the majority of painting techniques in use at the time. The individual volumes analyse, in a systematic fashion which has its precedents and will have its successors, both considerations pertaining to the composition and quality of the materials and the methods used to apply colour and make paintings.

Dardo Battaglini's little handbook, *Il consigliere dell'artista. Raccolta di ricette e consigli pratici indispensabili a tutti gli artisti* (Alessandria 1926), inherits the long tradition of formularies, listing technical entries in alphabetical order.

Giorgio de Chirico wrote *Piccolo trattato di tecnica pittorica* (Milan 1928), a precious manual which discusses not only the technique but the theory of art. Maria Bazzi acknowledges the work's merit of revealing how De Chirico worked and suggests that the state of conservation of the artist's paintings reveals the value of his precepts.[7]

Carlo Linzi, painter, student of the chemistry of colour and instructor at the Royal Academy in Munich, wrote a report in 1893 for the International Painting Congress in Munich. The work was later published in only a few copies in Treviso in 1895, and then republished by Hoepli in its series of "Manuals" under the title *Tecnica della pittura e dei colori (l'arte del dipingere ad olio) secondo Raffaello, Tiziano, Giorgione, Tintoretto* (Milan 1930). The rather original theories of this painter, who decorated Fogazzaro's "Montanina" villa and was celebrated by him in his novel *Leila*, involve use of asphalt or tar - historically unjustified, as noted by Maria Bazzi, who speaks of "personal theories and concepts which do not reflect historical tradition."[8]

Gaetano Previati combined his work as an artist, the role in which most people are familiar with him, with intense activity writing treatises. After *Memorie sulla tecnica dei dipinti* (1896), he published *La tecnica della pittura* (Turin 1905), *Principi scientifici del divisionismo* (Turin 1906) and *Della pittura, tecnica e arte* (Turin 1913).

As Silvia Bordini notes, "Previati thus fits into the centuries-old tradition of writings on technique, updating it with the most recent contributions, with the stated intent of helping painters choose their materials and proceedings; but he abandons the traditional form of the manual" to put "art and technique on the same level, one of profound interaction". Maria Bazzi suggests that *La tecnica della pittura* may derive from the precepts of Vibert,[9] undoubtedly identifying one of the painter and writer's sources.

Giuseppe Ronchetti of Bergamo authored a well-known and successful series of manuals published in the Hoepli series, including *Manuale per i dilettanti di pittura a olio, acquarello, miniatura, guazzo, tempera, encausto, pastello, fotopittura (paesaggio, figura, fiori, cartelli, reclame)* (Milan 1904), *Pittura murale - Fresco - tempera - stereocromia - pittura a olio - encausto* (Milan 1911), *L'arte di dipingere i fiori all'acquerello ad olio ed a guazzo sulle stoffe e specialmente sul ventaglio* (Milan 1908) and *La composizione delle tinte nella pittura a olio e ad acquerello* (Milan 1913). The author is particularly influenced by a didactic structure like that of French popular handbooks of the late nineteenth century. His works often include plates and illustrations.

Augusto Leone Rosa in *La tecnica della pittura dai tempi preistorici ad oggi* (Milan 1937) discusses his topics in a sporadic, rather unorganised fashion but supplies an incredible amount of information and formulae, often referring to ancient manuscripts such as, for example, the Volpato manuscript or those published by Merrifield.[10]

Francesco Vismara wrote *Manuale pratico di pittura* published by Sonzogno (Milan 1929), a small popular handbook complete with engravings and plates describing different painting techniques (oil painting, etching, gouache, pastel and fresco painting), preceded by a number of notions of drawing and perspective. In the preface to the work the author, addressing himself to the amateur and beginning artist, the primary target of his work, says "I flatter myself that I have written a little book that will be of use to those who do not disdain to add a knowledge of painting to their intellectual resources."

And finally, an essential work of historical assessment is that by restorer Tito Venturi-ni-Papari: *L'arte della pittura ad encausto e degli stucchi del tempo di Augusto* (Rome 1901).

Varnishes and Varnishing Products

Another field strictly related to that of artistic production is the sector of varnishing products and varnishes, almost always used as a finish for paintings but also as pigment binders and for covering surfaces made of wood, cardboard and other materials. In this genre of treatise we may often also find information on pigments, inks and so on.

Colori e vernici. Manuale a uso dei pittori, verniciatori, miniatori, ebanisti e fabbricanti di colori e vernici is a historic manual with nineteenth-century roots and style by Gemello Gorini, republished in 1896 for Hoepli by Giuseppe Appiani. The work was later replaced in the "Manuals" series by *Colori e vernici. Manuale a uso dei pittori, verniciatori, miniatori, ebanisti e fabbricanti di colori e vernici* (Milan 1912) by Max Mayer and Pietro Bonomi da Monte, respectively manager and director of Colorificio Italiano. This is one of the most interesting Italian manuals on varnishes and varnish paints, industrially made and otherwise, of particular interest in relation to its review of pigments, partly derived from the factory's repertory of supplies.

The many publishing initiatives of Raffaele Armando Califano-Mundo include a *Manuale delle vernici e pittura a vernici* describing varnishes made with soft resins on the basis of the traditional classifications, focusing prevalently on use in the fine arts. The part on varnish paints may be combined with the interest generated by the series of varnish colours produced by Fratelli Maimeri of Milan.

Ugo Fornari wrote *Vernici e prodotti affini. Lacche mastici* for Hoepli in 1892. It is an industrial reference for professionals in the fields of decoration and colour mixing, republished repeatedly in the first two decades of the twentieth century.

To this we may add *Industria delle vernici* (Milan, undated) by S. Fachini, a work of an industrial nature; *L'industria dei colori e delle vernici* (Milan 1916) by Ettore Rizzini; *Gomme, resine, gommo-resine e balsami. Origine, produzione, composizione e usi industriali dei principali prodotti vegetali di secrezione* (Milan 1909) by Luigi Settimj, containing a commercial description of the principal ingredients in varnishes used by craftsmen, paint makers and mixers and industrial paint manufactures; *Manuale teorico pratico del verniciatore* (Milan, undated) by Giovanni Sommaruga, a work intended primarily for professional varnishers including an interesting section on the composition of the hues of the day for decorators, and *Il verniciatore* (Florence 1913), a brief anonymous manual by a varnishing and decorating technician.

Decorative Painting and Gilding

A volume of great interest for historical as well as practical purposes is the painter-decorator's compendium or manual put together by Damaso Frazzoni, *L'imbianchino decoratore-stuccatore. Delle varie tecniche, antiche e moderne, per coloriture edili - modo di pulire i muri e patinarli all'antico - intonazione dei colori - superficie liscia e superficie lucida - doratura - norme igieniche* (Milan 1911). The work provides, by compilation but also and above all by collection of oral and practical traditions, a complete overview of the entire range of decorating techniques, particularly in architecture and wall decoration. It is extremely useful for its overview of the study of architectural finishes, as well as its collection of recipes which may be taken from manuals and formularies such as those by Italo Ghersi.

Another work of interest on the topic of architectural decoration is by Felice Zorza, *Il decoratore d'appartamenti. Manuale pratico ad uso degli apprendisti e dei dilettanti* (Turin, undated) and in the area of decoration of furnishing accessories, we may refer to Raffaele Armando Califano-Mundo's *Manuale per arazzi, gobelins, tele di Francia, da tappezzerie a secco, broccardi, treilles, fougère*.

Architect, historian and architecture critic Alfredo Melani, director of the Civic School at Castello Sforzesco in Milan from 1898 to 1923, wrote two interesting manuals on the decorative arts published by Hoepli: *Manuale d'arte decorativa antica e moderna* (Milan 1907), prevalently intended for use in ornamentation and decoration, and *Manuale dell'Ornatista* (Milan 1896), containing a list of suggestions and plates providing models for the decorator. Another work of great interest is the *Dizionario dell'arte e delle industrie artistiche illustrato. Monumenti e artisti italiani e esteri, tecniche antiche e moderne, stili e scuole d'arte* (Milan 1930), one of Melani's most important works, a sort of encyclopaedia in which the entries include technical notes.

The first half of the twentieth century saw the publication of a number of manuals concerned with gilding and metallisation techniques. These include *Indoratura, inargentatura e metodi di metallizzazione* (Milan 1895), an anonymous professional manual concerned primarily with electroplating, and T. Turco, *Tutti i sistemi di doratura. Manuale teorico-pratico ad uso degli artigiani e delle piccole industrie* (Turin, undated), one of the lesser known works of the indefatigable author of certain famous Hoepli Manuals published after World War II.

Though it may not be considered strictly a manual of the decorative arts, I also wish to list *La xilografia* (Milan 1926) by Giulio Cisari, a rare work on woodcutting written with admirable balance. The description of the proceedings involved in one of the most ancient engraving techniques includes a historical introduction which goes into significant depth for a small popular manual.

Drawing

It is very difficult to collect and classify works and manuals on drawing, in which considerations on artistic drawing are often mixed up with technical drawing or industrial drawing. There is as of yet no complete list of manuals and textbooks, particularly those regarding artistic anatomy, decoration and perspective, linked with teachings in academies, art schools and vocational institutes, often drawing on a vast repertory of models and descriptive plates.

Isidoro Andreani wrote numerous works published by Hoepli, including several volumes dedicated to the professions in the field of construction, as well as a complete drawing course: *Corso completo di disegno* (Milan 1909).

Well-known drawing manuals include *I principi del disegno e gli stili dell'ornamento* (Milano 1882) by architect and writer Camillo Boito. One of the bestselling Hoepli Manuals, it was republished fully seven times between 1882 and 1925. The work, intended for artists and craftsmen, is complete with an appendix, *Notizie di alcuni libri concernenti l'insegnamento elementare del disegno e l'arte ornamentale,* a sort of guidebook for orienting the reader in the most common illustrated repertories.

Teoria delle ombre con un cenno sul chiaroscuro e sul colore dei corpi (Milan 1897) by Elia Bonci, professor of drawing and painting, is one of the best manuals on the theory of shadows. Complete with plates, it was published five times in the Hoepli Manuals series.

Aldo Nanni attempts in *Il libro dell'allievo artista: il disegno dal vero senza maestro* (Turin 1932) to oppose the books of models and "teach drawing from life without a teacher", proposing his own original method.

The many manuals published by Hoepli written by Giuseppe Ronchetti, an author we have already discussed, include *Grammatica del disegno. Metodo pratico per imparare il disegno* (Milan, 1902, 2 voll.). This work, which was reprinted several times, offers an interesting blend of general information and sequences of repertory plates from the books of models. Numerous works were published specifically concerning anatomy and perspective. These include *Trattato di anatomia* by Raffaele Armando Califano-Mundo, who also wrote *Manuale del disegno e precetti*, an academic treatise offering a wealth of practical information; *Guida allo studio della anatomia artistica* (Milan 1905) by Giulio Valenti; *Manuale di prospettiva* (Milan 1903 - 2nd edition by engineer Claudio Claudi, a rigorous description of the theory of perspective; *Anatomia per gli artisti* (Faenza 1940) by Angelo and Giovanni Morelli, an agile manual for the teaching of anatomy intended for use in art schools and academies. Lastly, an anatomy professor from the fine arts academy in Carrara, Achille Lombardini, wrote two manuals on both these topics: *Manuale di anatomia pittorica* (Milan 1886), the third edition of which, dated 1907, was edited by Vittorio Lombardini, and *Il segreto della prospettiva: volgarizzazione alla portata di tutti della tecnica prospettica* (Turin 1935).

Restoration Techniques

In 1866 illustrious restorer Giovanni Secco Suardo published his famous *Manuale ragionato per la parte meccanica dell'arte del ristauratore di dipinti*, which met with considerable success and was considered an indispensible tool for restorers and collectors. The work was published by Hoepli in two volumes in 1894 under the title *Il ristauratore di dipinti*, and was reprinted several times (1918, 1927, 1979, 1988, 1993). Gaetano Previati wrote an *Introduzione allo studio del restauro*, included in the third edition of Secco Suardo's *Manual* (1918), in which he expresses his opinion on the restoration of ancient paintings.[11]

A pioneering work of the day was *L'esame scientifico delle opere d'arte ed il loro restauro* (Milan 1936) by Renato Mancia, director of the scientific research laboratory at the National Restoration School, opened in 1935: the first in Italy to use scientific aids to analyse paintings. In the preface to the work we may read: "What little has been said and written on restoration has now become obsolete; the new technical means, the products of the colorants industry, have wrought profound changes and improvements. The sciences of physics and chemistry have revolutionised the old methods of study."

An interesting document/brochure is a reprint or copy, without any date or place of publication but certainly dating back to the early twentieth century, of *Il restauro e la rigenerazione dei dipinti a olio di Massimiliano De Pettenkofer* by Uberto Giuseppe Valentinis (Udine 1874), discussing the controversial method of regenerating varnishes in paintings with alcohol vapours, and perfectly faithful to the original edition.

Lastly, as part of his intense publishing career, Raffaele Armando Califano-Mundo dedicated a work to restoration: *Alterazioni fisio-chimiche dei dipinti e norme generali per ovviarle.*

This is intended to be the first list of the principal published sources of information on technique in the arts in the first forty years of the twentieth century, but by no means claims to be a complete list.

This kind of text has largely been a niche area in publishing, which is why it is so difficult today to trace works which were undoubtedly famous and successful in their day, but were published by small publishing houses or without any indication of the date or author. The oldest publications and those of a commercial nature were often published by local printers who cannot easily be identified with the usual tools of bibliographic research.[12]

The initial results of the study are encouraging, but how many more manuscripts and typescripts still remain in archives and in the cellars of libraries, never having seen the light of day? How many turn-of-the-century catalogues, price lists and brochures abounding in technical information? How many folders and collections of notes, speeches and reports are kept in the bowels of the fine arts academies? How much correspondence among artists or between artists and suppliers, discussing not only theoretical questions but technical issues, remain unknown to us?

A few examples may give us an idea of the answer. The State Archive in Milan has a vast collection of artists' letters, definitely containing technical information but remaining entirely to be discovered.[13] The City of Milan's Bertarelli collection of prints still includes an original typescript, preparatory to publication of a book by Ugo Zovetti regarding paper marbling and decoration techniques; a valuable document by one of the most significant and "technical" Italian artists, an instructor at the Società Umanitaria, ISIA in Monza and the School of Applied Arts in Castello Sforzesco.[14] The manuscripts in the collection of the Angelo Mai Library include a surviving series of autographed copies, of considerable extent, attributed to Giovanni Secco Suardo, the best-known late nineteenth century Italian restorer and student of artistic techniques. The very folder containing this *Treatise on Paint* includes another work of considerable significance, also by Maimeri: a *Study of Colour* offering a wealth of important links with phenomenological philosophy.

Unfortunately, historical and artistic study, even of the modern and contemporary ages, is still hindered by centuries-old methodological impasses. Only recently, and with difficulty, has the study of the history of artistic technique been granted full recognition as a historical and artistic discipline, and, we hope, this comes at the end of a "not so brief and insignificant time of historiographic orientation which tended to undervalue or exclude from the critical eye assessment of aspects pertaining to materials and proceedings, which were more or less explicitly considered insignificant or of lesser importance. These beliefs were rooted in the ancient conception of art as a servile task, linked with the handling of materials and the use of tools, and were strengthened in the course of the debate sanctioning the artists' emancipation from the manual sphere and inclusion in the intellectual sphere of the liberal arts, and were finally radicalised in the idealistic attitude that viewed art as an activity of the spirit".[15]

I realise that archivistic and bibliographic research work of this kind cannot lead to immediate availability of certain information, but will uncover partial information, to be assessed with great care and precaution in historical and artistic terms. But, as Francesca Tolaini rightly notes, in the proceedings of study, if we take into account the partial nature of the research, it may in any case be useful to support more general observations of use for overall assessment of artistic actions which would otherwise be "even less reliable".[16]

[1] S. BORDINI, *Materia e immagine. Fonti sulle tecniche della pittura*, Rome 1991.
[2] F. TOLAINI, "Breve storia dello studio dei ricettari di tecniche artistiche medievali", in *Annali della Scuola Normale Superiore di Pisa*, series IV, vol. VI, 1, Pisa 2001, pp. 11-38 (p. 12).
[3] M. BAZZI, *Abecedario pittorico,* Milan 1956 (English ed. *The Artist's Methods and Materials*, London 1960).
[4] I. GHERSI, *Ricettario domestico,* Milan 1899 (3rd ed. 1906, revised by Arturo Castoldi); I. GHERSI, *Ricettario industriale,* Milan 1899.
[5] L. SOLARI, *1300 ricette utili per la casa. Il libro dell'autarchia domestica*, Turin 1943.
[6] V.G. VILLAVECCHIA, G. EIGENMANN, *Dizionario di merceologia e chimica applicata*, VIII, Milan 1973.
[7] BAZZI 1956, p. 307.
[8] BAZZI 1956, p. 309.
[9] BAZZI 1956, p. 311.
[10] M.P. MERRIFIELD, *Original Treatises dating from the XIIth to XVIIIth centuries on the Arts of Painting...*, London 1849 (repr. *Medieval and Renaissance Treatises on the Arts of Painting*, New York 1967 and 1999).
[11] In this regard cf. S. BORDINI, "L'Introduzione di Gaetano Previati a Il Ristauratore di dipinti di Giovanni Secco Suardo", in *Giovanni Secco Suardo. La cultura del restauro tra tutela e conservazione dell'opera d'arte*, acts of the International Study Convention (Bergamo, 9-11 March 1995), supplement to issue 98 (1996) of the *Bollettino d'Arte*, 1998, pp. 115-118.?
[12] Cf., for example, the interesting brochure *Istruzioni per la tempera De Pereira*, distributed by Kaspar & Dr. Vogl (the first Austrian company to manufacture paint sold in tubes) and printed in Italy by M. Fabricatore (Naples 1910). Il Fabricatore was the printer-publisher who published the first works by Califano-Mundo. The brochure in question, with its own separate numbering, was annexed and stitched to the Neapolitan printer's publications and, at the same time, circulated independently in paint manufactures and shops in Rome and Naples. It is possible that the anonymous writer of the text may have been Califano-Mundo himself, or at least that he played an important role in the revision of the commercial brochure. For he had in fact set up a special technical office for the analysis of products for the fine arts which were sent to him by the manufacturers. The office, having verified the quality of the products, reviewed them and publicised them in the director's publications. This made for an even closer relationship between Califano-Mundo's complex publishing career and the productive plant and business operated at the same time.
[13] I observed the collection during PROMATART, a project for surveying, cataloguing and studying technical and artistic manuscripts kept in the main libraries of Lombardy, directed by Sandro Baroni and financed by Fondazione Maimeri and the Region of Lombardy. Cf. G. CAPROTTI, P. TRAVAGLIO, *Presentazione dei risultati del progetto Promatart*, convention on the topic of "Illuminated manuscripts: where art and science meet", on the first national day for the multidisciplinary study of illuminated manuscripts, Parma, 13 June 2008.
[14] In this regard, cf. S. BARONI, *Manuel pratique de l'imitation du marbre*, Paris 1996, pp. 99-101; S. BARONI, *Decorare la carta. Corso pratico*, Milan 1998, p. 42.; A. MILANO, E. VILLANI, *Le carte decorate della Raccolta Bertarelli*, Milan 1991.
[15] BORDINI 1991, pp. 8-9.
[16] TOLAINI 2001, p. 38.

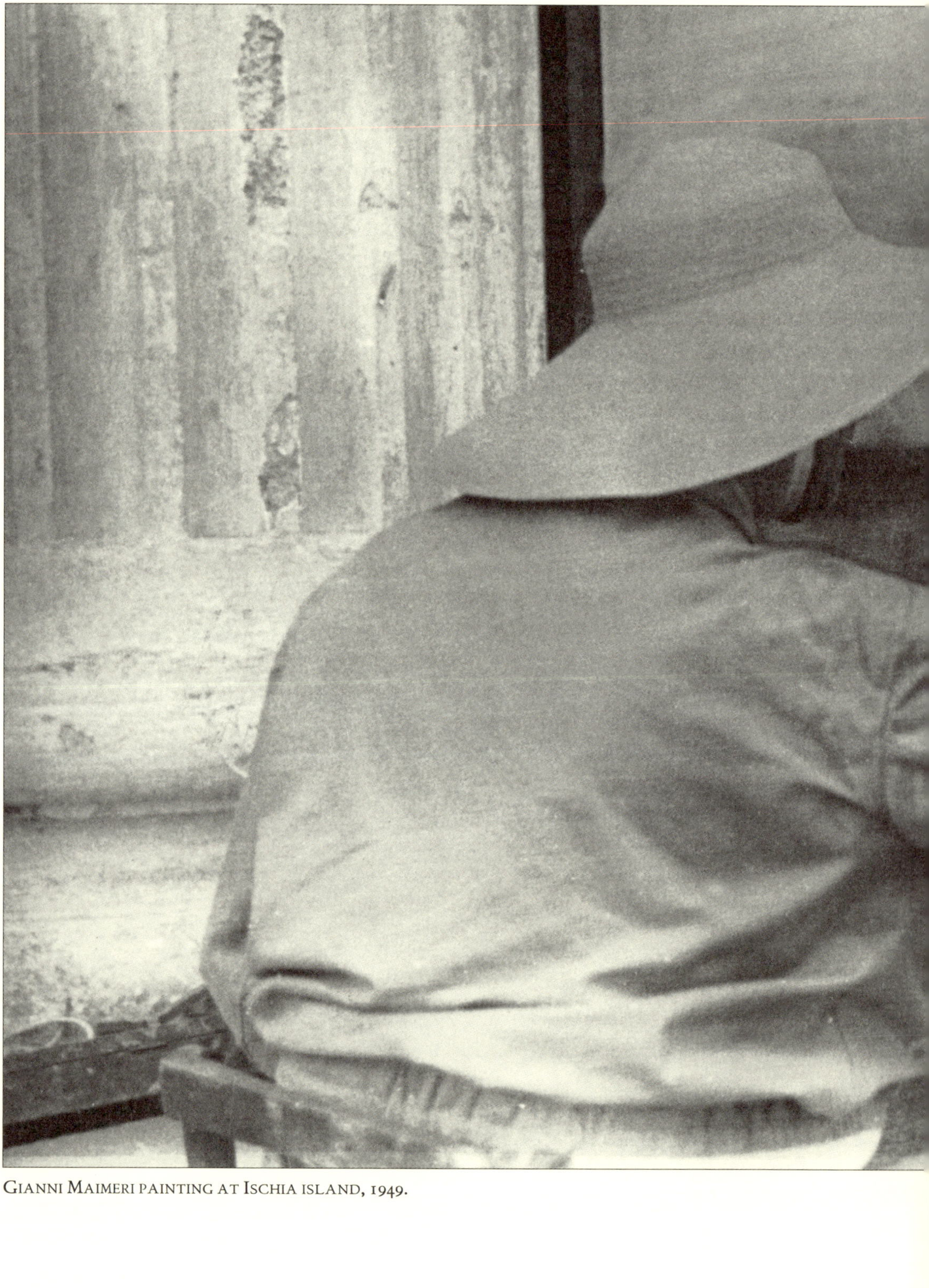

Gianni Maimeri painting at Ischia island, 1949.

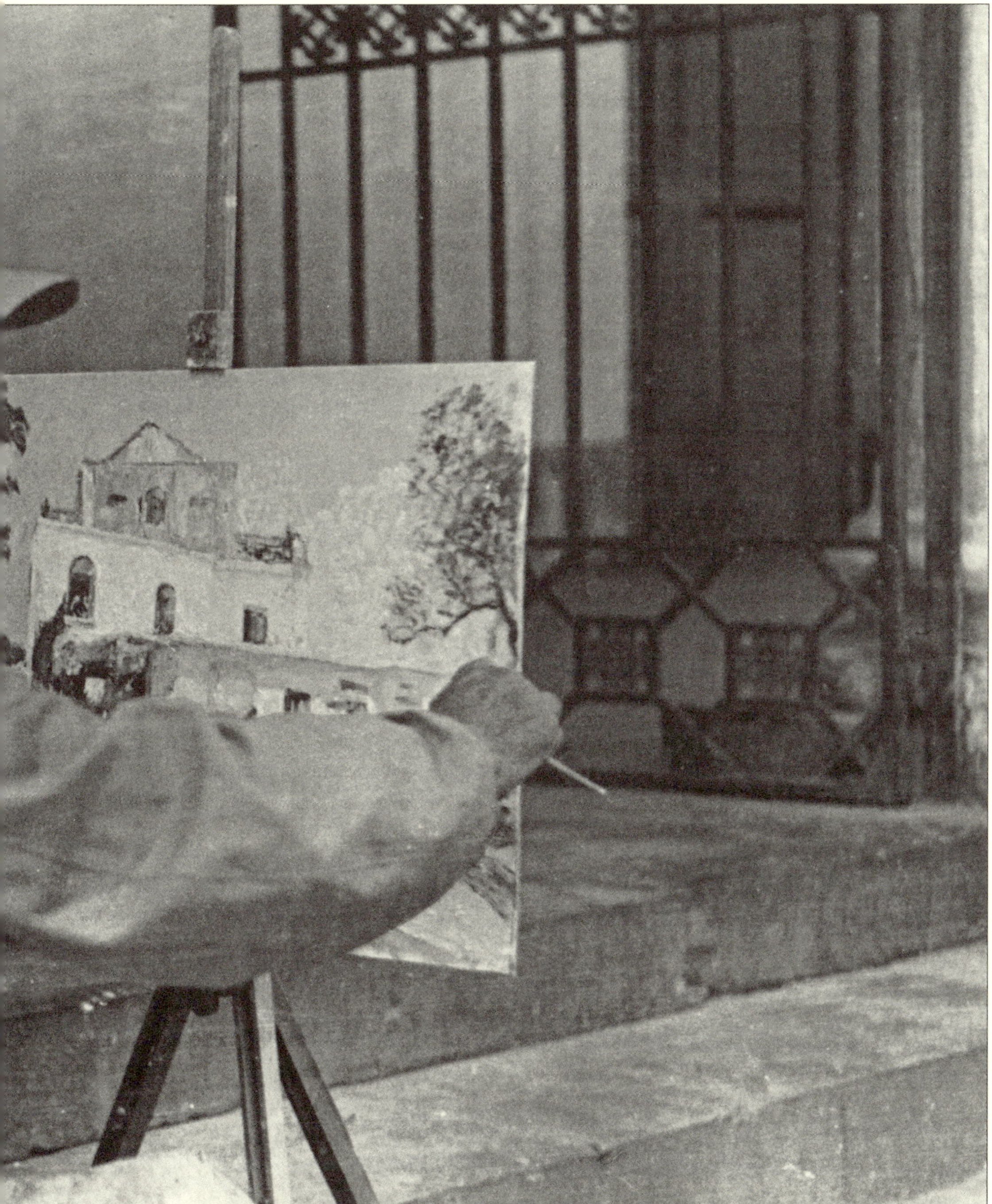

GIANNI MAIMERI, *Treatise on Paint*
Edited by Sandro Baroni

Contributions by
Paola Travaglio
Luigi Brusati
Carlo Migliavacca
Maite Rossi

Translations
Darin Communication

Graphic design, impagination and printing
Umberto Allemandi & C

Special thanks to Fondazione Maimeri
for consenting to publication of texts and images
from the Foundation's historical archive.

First edition, January 2010

COPY-EDITING KARIN JUDKINS
DESKTOP PUBLISHING ALESSANDRA BARRA

PHOTOLITHOGRAPHY FOTOMEC, TURIN

PRINTED IN THE MONTH OF JANUARY 2010
BY TIPO STAMPA, MONCALIERI (TURIN)